The Biggle Sheep Book
Something Practical About Sheep

by Jacob Biggle

with an introduction by Jackson Chambers

Self Reliance Books

Get more historic titles on animal and stock breeding, gardening and old fashioned skills by visiting us at:

http://selfreliancebooks.blogspot.com/

Introduction

I am pleased to present another book in the famous "Biggle Farm Book" series.

The work is in the Public Domain and is re-printed here in accordance with Federal Laws.

As with all reprinted books of this age that are intended to perfectly reproduce the original edition, considerable pains and effort had to be undertaken to correct fading and sometimes outright damage to existing proofs of this title. At times, this task is quite monumental, requiring an almost total "rebuilding" of some pages from digital proofs of multiple copies. Despite this, imperfections still sometimes exist in the final proof and may detract from the visual appearance of the text.

I hope you enjoy reading this book as much as I enjoyed re-publishing and making it available to farmers, homesteaders and back-to-the -landers again.

With Regards,

Jackson Chambers

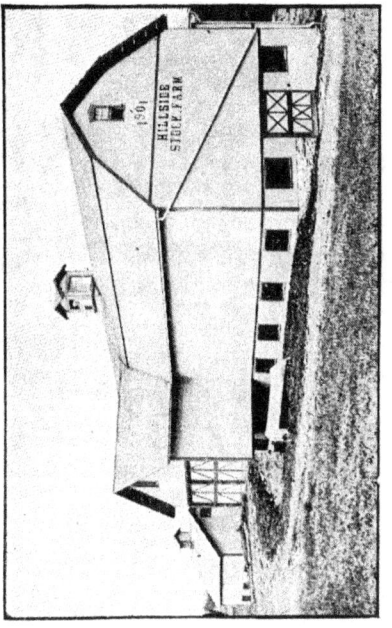

Fig. 11—C. E. COLBURN'S FARM AND STOCK BARN

1

Fig. 19.—MR. LAWSON VALENTINE'S BARN, "HOUGHTON FARM," MOUNTAINVILLE, N. Y.

A PRETTY SIGHT. OVER FOUR HUNDRED, LITTLE AND BIG.

CONTENTS.

BREEDS ILLUSTRATED.

The representatives of the English mutton breeds, and of the Rambouillets, are photographs from life of the champions at the International Live Stock Exposition held at Chicago, which is conceded the greatest annual live stock show held. The champions in such a show are typical representatives of their respective breeds. The other photographs are typical specimens of their breeds, being noted animals.

PREFACE.

The sheep industry of the United States is divided geographically and every other way by the Missouri River, and the industry as carried on west of this line is quite different from that east of it. Figured in number of sheep and capital invested the West forms the larger half of the business, but reckoned by the number of individuals engaged in sheep raising the East predominates. Obviously in a condensed work like this it would hardly be possible to treat at all adequately the industry of both sections, so that governed by the consideration of the greatest good to the greatest number, I shall confine myself mainly to the sheep business as commonly carried on in connection with other lines of agriculture.

To be brief and to be practical are the chief aims of this work. It is intended not as a scientific manual for the educated theorist, not as a complete and authoritative work on sheep for the experienced shepherd, nor yet as a kindergarten primer for some theorist or faddist who may happen to turn his attention to sheep as a fancy, but it is intended to be a practical guide to the average farmer who tills the soil, and who, wishing to advance and progress with the times, keeps a flock of sheep to improve his soil, increase his crops and wealth and add to his general welfare.

I shall not attempt fine theories or impracticable methods, simply the means to be used in the every-day care of a flock of sheep. Whatever finds expression in these pages is gleaned from experience.

Among others to whom I wish to express my thanks in aiding me in preparing this brief handbook and guide, I wish to mention W. J. Clarke, " Shepherd Boy," as he is known to the readers of current sheep lore ; to the secretaries of the various sheep breeders' associations for information in regard to the various breeds ; to Messrs. C. S. and Max Chapman, of Ohio ; F. B. Hartman, of Indiana ; H. E. Moore, of Michigan, and others for photographs. To A. A. Wood, of Saline, Michigan, upon whose inexhaustible fund of practical knowledge of sheep and all that pertains to them, secured by the wide experience of a lifetime, I have made large drafts, and to R. M. Wood, son of the last named, who is also a storehouse of practical and expert information about sheep.

CHEVIOT EWES READY FOR SUMMER.

PLATE I.

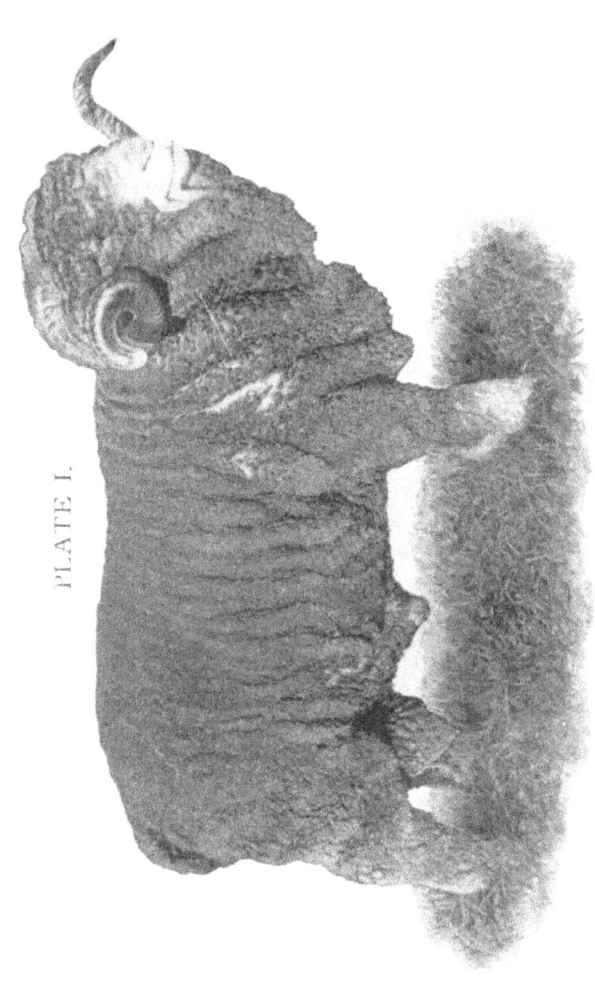

AMERICAN MERINO RAM.   FOUR YEARS OLD.

Chapter I.

HISTORY.

 From the beginning of history and through all the ages the sheep has ever remained man's quiet and faithful friend and helper, and to-day, with its golden hoof, it is the boon and blessing of the tiller of the soil.

A source of pleasure and profit to the patriarchs of sacred history, ever holding high position in the economic welfare of the small industrial world of the cultured Greek and conquering Roman, sheep were improved and increased ; throughout the dark and troublous times of the Middle Ages the industry was preserved, and during the rapid rise and advance of manufactures and all industrial pursuits of the present, sheep have kept pace with the march of progress and improvement, and still hold firm their position in the welfare and prosperity of the farmer.

With the first comers to the New World came the sheep : with Columbus in 1493 came sheep from the flocks of Spain, both coarse wool and fine Merino ; with the English gentlemen to Virginia in 1609 sheep from English flocks, with wool of medium fineness ; with the Puritan fathers in 1624 to 1629 came the large, fine-wooled Wiltshire sheep from England ;

with the Dutch to New York in 1625, the long-legged, coarse-wooled sheep of Zealand and Texel in Holland ; in fact, with all the colonists came sheep from the native land of their owners.

These colonial sheep sheared from two to three pounds of medium to coarse wool, and weighed forty-five to sixty pounds, and from them descended the present native stock: those of the Eastern States from the sheep of the thirteen colonies ; those of the West from the old Mexican ewe, the descendant of the Spanish importations of Columbus and the Spanish explorers.

Later, in 1801-11, came the Spanish Merino from Spain, with its remarkably fine fleece. These were widely disseminated, and improved the native flocks. In 1850-60 came many of the English mutton breeds, but until the last two decades wool has been the first, and practically the only consideration of the American sheep owner.

But during this latter time has arisen and grown an ever increasing demand for mutton, and with it came large importations of all the English mutton breeds, and a corresponding decadence in the use of Merinos. To-day the sheep business is in a somewhat unsettled condition, with mutton the important factor with some sheep growers and wool with others.

While local conditions will require special purposes, the average sheep owner will ever aim toward the best combination sheep producing both wool and mutton.

CHAPTER II

STATISTICS.

Let us have fewer dogs and more sheep.

According to the census of 1910, there were in the United States 29,707,000 ewes, 7,148,366 rams and wethers, 12,168,278 lambs, a grand total of over fifty-one million sheep and lambs. Of these 20,584,644 are east of the Missouri River ; only twenty million sheep on eastern farms where there should be fifty million.

Of all the states Montana has the largest number, with 5,372,639 ; Wyoming comes next with 5,194,959 ; in both of these states sheep are handled in large bands under range conditions. Of the eastern states Ohio has the most, with 3,907,055, and Michigan next, with 2,306,476.

Of all this number, not over half a million are pure-bred registered sheep ; about ten million are grades and cross-breds, that is, more than half pure-bred, while the great majority of four to one is native, or scrub, and there are many of the latter that are everything that the word implies.

The average value of all sheep in the United States may be estimated at from three to four dollars per head, according to the time of year. The total valuation is estimated at one hundred and eighty million dollars.

Chicago is the greatest sheep market in the world, 5,229,294 sheep, valued at $29,346,532 being handled there in 1910; the largest day's receipts were 63,000; the largest for a week 200,000. During 1910 Omaha handled 2,984,870; Kansas City, 1,841,173; Buffalo, 1,389,000. Ten leading markets handled a grand total of 14,514,280 sheep in 1910, while there is no means of estimating the number marketed and slaughtered by the growers and local dealers throughout the country.

In the last decade a large movement of feeders from the markets to farmers in the country has developed. In 1910 nearly three million feeders were sent from three markets in four months.

The estimated total wool clip of 1910 was 336,896,903 pounds, at an average valuation of 22 cents per pound. The estimated average shrinkage is sixty per cent. The average weight of fleece is approximately seven pounds.

The largest wool shipping point from the grower is Miles City, Montana, which shipped about ten million pounds in 1910; Great Falls, Montana, shipped about eight million pounds, and Casper, Wyoming, about six million pounds.

The total importations for 1910 were about one hundred million pounds, and exports practically none.

Every person in the United States wears out seven pounds of wool every year, on an average.

The heaviest sheep on record is a Lincoln ram, weighing 456 pounds; the next heaviest is an Oxford, weighing 436 pounds.

The heaviest fleece on record is that of an American Merino, weighing 52 pounds; followed by another Merino, with 44¼ pounds at public shearing.

PLATE II.

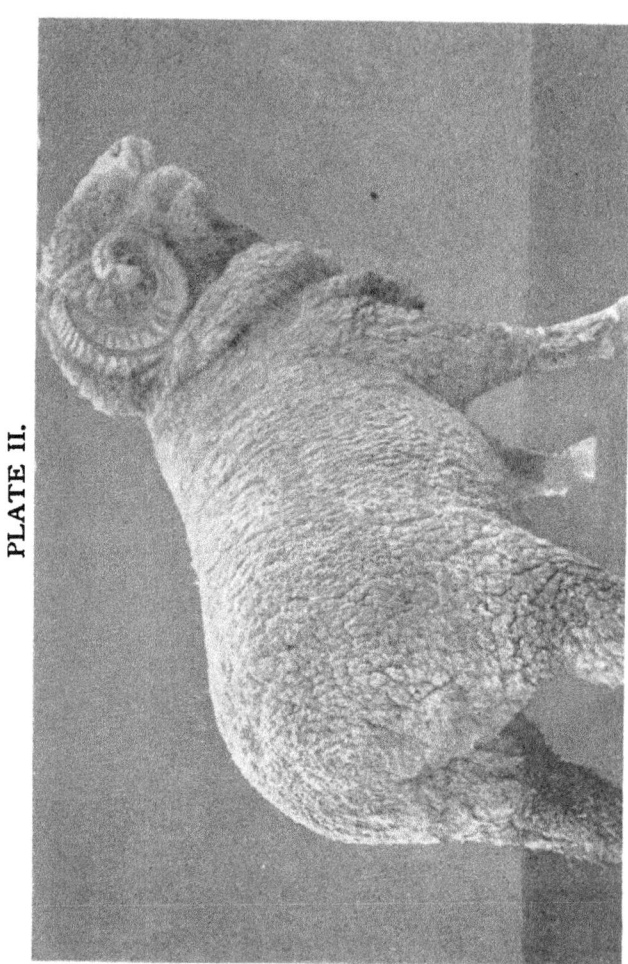

CHAMPION DELAINE MERINO RAM. TWO YEARS.

Chapter III

BREEDS.

That breed is best which will receive the best care.

I must necessarily be confined to only the most essential and general points of the various breeds, and do not aim to give preference to one breed over another; there is room for all, and all are profitable in their place, except the scrub, which has no place in the flock of any farmer who expects to succeed. A list of the secretaries of the various breed associations is given in the back of the book, any one of whom will cheerfully give fuller information concerning his particular breed.

The Merino is a native of Spain, the Spanish sheep in turn being descended from the famous fine-wooled sheep of Tarentum, in Italy. The ancestors of the Merinos in America were exported from 1783 to 1811. They are hardy, long-lived, small to medium size, unequalled for fineness and weight of fleece; native to a cool, dry climate and a broken and somewhat mountainous country; very adaptable to changing conditions of climate and locality, to running in large flocks and thriving on short, scanty herbage.

The Spanish or American Merino.—(See Plate I.) Descended from importations made to the United States, direct from Spain, in 1799 and the succeeding years to 1811. First flock record established in 1876. First radical development was toward wrinkly, oily, heavy-fleeced sheep; some breeders continue to breed this style of sheep, density and

weight of fleece being their object. These extremely wrinkly Merinos are now known as A type Merinos, or Vermont Merinos. Later and present tendency of the majority of breeders is toward larger, smoother sheep, with heavy neck and hip folds, but plain bodies, longer staple of wool and freer from grease.

AMERICAN MERINO RAM.

These are now known as B type Merinos. Average weight of rams, 150 to 175 pounds, and shear 23 to 35 pounds; ewes weigh 115 to 130 pounds, and shear 15 to 22 pounds. Wool grades heavy fine, fine, and fine delaine, with a staple of 2 to 3 inches. They are essentially a wool sheep, very long-lived, as well as having all other qualities of a Merino.

DELAINE.—(See Plate II.) Same origin as American, but aim in breeding has always been toward a long, fine staple of wool, free from grease, and more attention to the form of the sheep. They were first developed in Pennsylvania and Ohio, solely by selection, breeding and feeding. There are several different families and registries as the Standard, the Black Top, the Improved and the Dickinson, these differences being mainly due to blood lines in breeding and some minor characteristics, but all are essentially Delaines, aiming toward the same kind of sheep. Rams weigh 160 to 200 pounds, and shear 18 to 25 pounds; ewes weigh 120 to 140 pounds, and shear 12 to 16 pounds. Wool grades fine delaine, with a staple of 3 to 5 inches.

RAMBOUILLET.—(See Plate III.) Origin, Spain; imported to royal estate at Rambouillet, France, in 1786, whence the name. Imported to the United States in 1846, 1851, and also recently. Bred for a combination of wool and mutton, with the idea of producing a sheep with a good fleece and the essential Merino characteristics and at the same time a good marketable mutton body. Flock record established in 1891. Rams weigh

CHAMPION RAMBOUILLET EWE "BERNICE."

225 to 300 pounds, and shear 18 to 25 pounds; ewes weigh 150 to 200 pounds, and shear 12 to 16 pounds. Wool grades fine to fine medium, with a staple of 2½ to 4 inches. They are the largest of the Merinos, and are very popular both on eastern farms and western ranges.

It is claimed that the Rambouillet is easily kept and that it is peculiarly well adapted to the range country. The breed is dense-wooled, large and smooth, and produces large lambs, running in weight from 80 to over 100 pounds when 6 months old.

MUTTON BREEDS. — Native to Great Britain; quick maturers, short-lived, comparatively light fleeced, and generally adaptable to intensive farming and artificial conditions of feed, primarily for mutton. Wool grades medium to combing.

HAMPSHIRE-DOWN.—(See Plate V.) Natives of the ridge lands of the chalk districts lying south of London, England. The district of Hampshire is their home. From the best authority attainable they undoubtedly originated from crossing Southdown rams upon the native black-faced sheep of Hampshire and Berkshire. A few flocks were started in America in the South before the war, but became entirely annihilated during the progress of that episode. The first organized effort for their importation to this country of which we have record was in about 1875. Flock record established in America in 1889. Mature rams weigh 250 to 350 pounds ; ewes weigh 150 to 250 pounds. Flocks shear from 8 to 12 pounds. Wool is known as three-eighths blood ; it is a fine quality of clothing. They will thrive in any climate, on any soil, and under any conditions that any sheep will ; a dry soil and temperate zone preferable for these, as well as for all other sheep. Excellence : early maturity, constitutional vigor ; docility and motherly qualities ; adaptability ; fecundity ; prepotency. The Hampshire is a very large sheep, and has given excellent satisfaction as a producer of feeding lambs for the early and Christmas markets.

OXFORD-DOWN. — (See Plate VI.) Natives of Oxfordshire, England. Originated from crossing Cotswold rams on Hampshire ewes. First importation to the United States in 1846, again in 1853, to Delaware and Virginia. Flock record established in 1904. Rams weigh 250 to 350 pounds, and shear 12 to

16 pounds ; ewes weigh 180 to 275 pounds, and shear 10 to 12 pounds. Wool grades medium. Excellence : weight of the Longwool, quality of the Down.

They are the largest of the Down breeds, and have the heaviest and coarsest fleece. They are best

A NICE FLOCK OF OXFORD-DOWNS.

adapted to intensive farming, and will thrive on abundant pasture, which inclines to grow somewhat coarse and rank.

SHROPSHIRE-DOWN.—(See Plate VII.) Indigenous to Shropshire county, England, known in the thirteenth century. First importation to the United States was in 1844. Mature rams weigh 225 pounds, and shear 12 to 16 pounds ; ewes weigh 165 pounds, and shear 8 to 12 pounds. Wool grades best *medium* delaine, with a staple of 4 to 6 inches. Excellence : early maturity, fecundity, prolificacy, style, appearance, docility; generally adaptable to farming conditions in the East, but do best where the pasture is abundant and the land comparatively level.

Of all the mutton breeds, Shropshires are the most popular and most extensively used in the United States and Canada ; they have more recorded animals than all the other mutton breeds combined, and have the largest sheep registry association in the United States. The cross of Shropshire rams on native ewes

has been very satisfactory in furnishing mutton lambs for feeding.

SOUTHDOWN.— (See Plate VIII.) Native to Sussex, England ; from the earliest known period they were found on the downs of Sussex, but were first improved by John Ellman, Glynde,

PRIZE YEARLING SHROPSHIRE RAM.

England, in 1780. The first importations into the United States were in 1824-8, to Pennsylvania and New York. Rams weigh about 200 pounds and shear 9 pounds; ewes weigh 150 pounds and shear 7 pounds. Wool grades fine medium, being the finest of the medium wools. They will thrive where climate, soil and conditions are unfavorable, but are best adapted to improved farming conditions. They are hardy, early maturers, produce a fine quality of meat, it being well graded with fat and lean, and is juicy and finely flavored. They have been bred pure for many years, and impress their good qualities upon the native stock ; they have contributed much to the origination of several of the other pure dark-faced breeds. They are the smallest of the English mutton breeds in America.

CHEVIOT.—(See Plate IX.) Native to the Cheviot Hills of Scotland ; there is a legend that they are descended from some sheep saved from the ships of an invading Spanish Armada that were wrecked off the coast of Scotland, but they have been in Scotland for many hundred years. The first importation to the United States was in 1842, to Otsego county, New York, where they are still very popular. Flock record established in 1891. Rams weigh 200 to 350 pounds,

and shear 12 pounds ; ewes weigh 150 to 200 pounds, and shear 7 to 9 pounds. Wool grades medium, with a staple of about 5 inches. They are adaptable to almost any condition, but especially to upland, dry pastures. They are hardy, free from disease, easy keepers, and large milkers.

DORSET-HORNS.—(See Plate X.) Native to Dorsetshire and Somersetshire, England. First imported to the United States in 1887 ; to Canada in 1885. Average weight of rams, 200 to 225 pounds, fleece 8 pounds ; ewes weigh 140 to 175 pounds, shear 6 pounds. Both rams and ewes have horns. Special excellence : winter, or hothouse lambs, as they are very prolific, often breeding twice a year.

LEICESTER.—(See Plate XI.) Native to Leicestershire, England ; they originated from the old long-wooled sheep of that county, which were called the Dishley breed. They were improved by the first improver of modern live stock, Robert Bakewell, by

selection and in-and-in breeding. First importation
to the United States was in 1812, by a Mr. Lax, of
Long Island, New York. Flock record established in
1888. Rams weigh 225 to 250 pounds, and shear about
12 pounds ; ewes weigh 175 to 200 pounds, and shear
9 to 10 pounds. Wool grades long or combing, with a
staple of 10 to 12 inches. They are adapted to almost
any soil or climate, with proper protection for winter,
but do best on arable land, and in a climate which
furnishes an abundance of succulent vegetation. Excellence : their size, and the ease with which they lay on
fat, also the fine bone and light offal, together with
length and weight of wool. They have been most
popular and widely used in Ontario, Canada ; they
have also been extensively used in improving the
other longwool breeds.

LINCOLN.—(See Plate XII.) Native to Lincolnshire, England. They were improved by selection

and careful breeding,
from the old native
sheep of Lincolnshire,
dating back about 150
years. The first im-
portation to the United
States was about 1872.
Flock record estab-
lished in 1891. Rams
weigh about 300
pounds, and shear 20 to

TWO BEAUTIES AND THEIR OWNER.

25 pounds ; ewes weigh
250 pounds and shear 15 to 18 pounds. Wool
grades combing, is long and lustrous, with a staple of

10 to 12 inches. They are adapted to almost any climate ; the best flocks in England are raised on both light and heavy soils, and the fens or lowlands of Lincolnshire, showing that they adapt themselves to any soil. Like all other sheep, they need a dry place to lie, and good feed ; they probably do best on level, arable land, where pasture is abundant. Excellence : hardy, good feeders, prolific, produce a large amount of mutton, and a long luster wool adapted to produce a certain kind of material that can be obtained from no other wool.

COTSWOLD.—(See Plate XIII.) Native to Gloucestershire and Oxford counties, England. The original Cotswolds were one of the earliest breeds of sheep in England, being noted for their wool production ; about the beginning of the 19th century, they were

A GOOD BUNCH OF COTSWOLDS.

improved by the crossing of Bakewell Leicester rams on the Cotswold flocks ; since then improvement has been by selection and care in breeding. The first importation into the United States was to New York in 1832. Flock record established 1878. Average weight of rams is 225 pounds, shearing 18 pounds; ewes weigh 175 pounds and shear 14 pounds. Wool grades coarse combing and braid, with a staple of 10 inches. They are best adapted to a limestone soil, and will adapt themselves to most parts of the United States and Canada, but require good pastures. Excellence : they combine mutton form, weight of carcass, with heavy fleece. This wool is always in quick demand, and close to the top of the market in price.

There are several other breeds in Great Britain which have not yet been generally introduced into the United States, so that they need not be considered here.

A WELSH MOUNTAIN RAM.

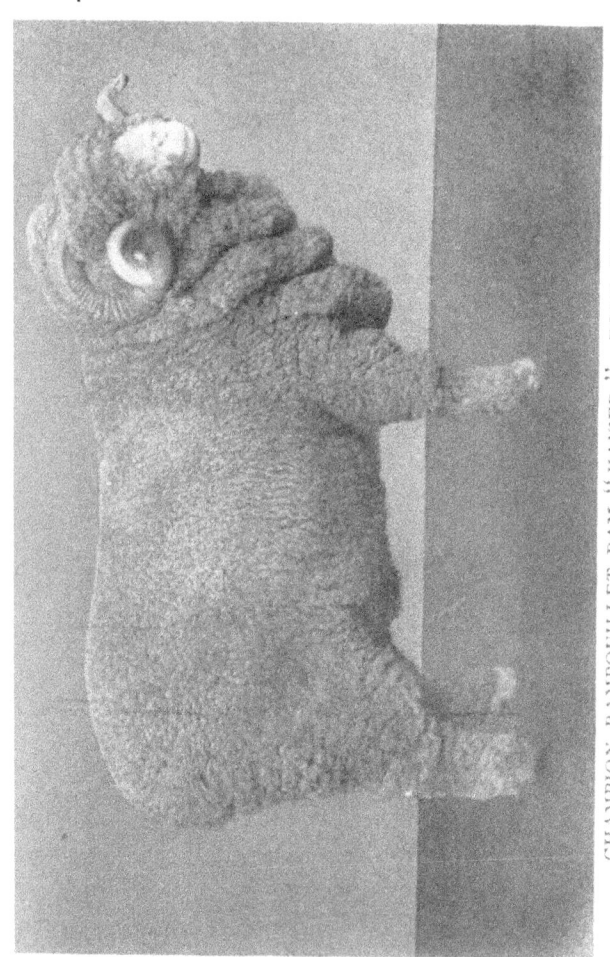

CHAMPION RAMBOUILLET RAM "KAISER." FOUR YEARS OLD.

CHAPTER IV

CLIMATE AND LOCATION.

Sheep are the most adaptable to change of climate and soil of any of the domestic animals.

 Different breeds are adapted to different conditions, but all sheep do best in a comparatively dry and temperate climate, on a dry soil, where the contour of the country is rolling and broken. The land must grow good grass, for that is the principal feed all the year, especially clover and blue-grass.

High, rolling land of clay loam, with limestone subsoil, is the best for sheep. It furnishes a great variety of grasses that are sweet and nutritious. A sandy soil does not furnish as good feed, but sheep will do well and improve the soil. In any case dry land is necessary ; sheep and water, so far as soil is concerned, do not make a profitable combination. In the semi-arid regions of the West, where grass is the only feed and snow furnishes nearly all the water, sheep have proven the most profitable of all industries.

Bottom lands are generally too wet for sheep, inducing worms and foot-rot, the two great enemies of sheep in the East.

Sheep will enrich a farm. No manure so strengthens and fertilizes the soil as sheep manure. A farm a little light or too heavy makes good sheep land, and sheep will improve it and increase its productive capacity.

Except for hothouse lambs, the location as regards markets for sheep raising is immaterial. Buffalo is the greatest sheep market for Eastern bred and fed stock, while all the other large cities have good stock yards and markets. There is always a good market for wool and mutton everywhere.

STEPPING STONES.

The worst enemy of the sheep is the dog, and the worst variety the bird dog.

Keep posted as to the markets, both wool and mutton; know what your products are worth.

There is no animal on the farm that returns so large profits for the money invested as a sheep.

Subscribe for up-to-date sheep and farm papers, and then read them; you can not learn too much.

A flock of the right kind of sheep is likely to be as solid personal property as the farmer can have.

In the case of Mutton vs. Pork it is a strong point that sheep never suffer from hog cholera or swine plague.

Of the domestic animals, the sheep furnishes the richest manure; it is better than any commercial fertilizer.

Sheep are good animals for hilly land. They do not require much of cultivated crops, and they do require dry footing.

There is a place for everything, but to succeed, everything must be in its place. Out of its place, the best thing is worthless.

Sheep are the best of manure spreaders, as well as manure makers; but they should be kept in the fields and not on the road, as we too often see them.

There is no use in keeping sheep if you haven't time to bother with them; but you can do nothing that will pay you better for the time and work spent than sheep.

CHAPTER V

THE RAM.

Don't pet or tease the ram; it will make him ugly and dangerous.

The ram is half the flock; yes, the whole flock, for without a good ram no breeder can produce good lambs. Always use pure bred rams. Go to any good reliable breeder of your breed of sheep, get a good ram, and do not be afraid to pay for him. The difference in the cost of a good sire and a poor ram is repaid many times by the increased value of the first crop of lambs.

Buy a ram that has individual merit and breeding to back it up. He must have a strong constitution, be active and vigorous; look well to the size of bone in the fore leg. His head should be masculine and denote strength, broad and of medium length; short, thick neck, set on full, straight shoulders, with broad chest and full heart girth; strong, broad back, round, well-sprung rib, with straight hind legs.

Look to size and form first, then to fleece. Have as thick a fleece as possible, with as much staple as the density will warrant; head, leg and belly should be well covered, and the fleece should be of even quality all over the body.

Masculinity and strength are the essentials. A good head, strong fore leg and full heart girth generally denote a good ram.

Do not buy a show ram; he may look nice, but one in good thriving condition will give you better results. Buy a well-bred ram, for good breeding is next in importance to good individuality. You are not buying a ram for the pounds of mutton or wool there is in his body, but it is what he will produce that you are buying. Get a better ram each time you buy.

The time to secure your ram is in the late summer or early fall. Do not wait until the day you want to use him, and then expect to find exactly what you need. Get your ram a month or more before you want to use him, so that he will get accustomed to you and his surroundings, feed, water, etc. August to October is the principal time to buy a ram.

FOOT NOTES.

Do not think that an animal must be imported to be good.

Early lambs sell the best, and oats will make them grow the fastest.

The stock ram should have breeding as well as individual merit, but a large part of it should show on his back.

The object of keeping and feeding animals is profit, and it should always be a matter of study how to get the most weight at the least cost.

To stock a certain amount of land with sheep requires less capital than to stock the same land with cattle or horses. This is a point for the man with limited means.

For a ram that is continually fighting or knocking the pen down, take a piece of cloth, soft leather or felt (the leg of an old felt boot will do), have it large enough to cover the face and eyes, tie it around his horns, and let him go.

SHEEP THRIVE EVEN ON ROCKS, GOOD FOR NOTHING ELSE.

Chapter VI.

THE EWE.

Don't let tne butcher coax away the best ewe lamb.

Every flockowner should start with a small flock; handle what he can take good care of, and increase the number with increasing accommodations and experience.

Having decided on the breed, breed your flock pure; there is nothing gained by crossing the breeds. Such produce always proves uncertain and unsatisfactory as breeders, and continual crossing soon results in a scrub. Have your flock uniform; they should look as near alike as possible.

See that they look like ewes and not like rams. Have them good average size, well proportioned and symmetrical. A feminine head, clean-cut nostrils, clear eye, small neck, long and thin as compared to a ram's, strong shoulders with good heart girth, round, well-sprung rib, back slightly arched, with well-filled hips and straight hind legs.

The ewes must be strong boned and have good constitutions; the size of the fore leg is a good indicator of both. Life and vigor in the ewes means strong lambs.

Look well to your fleeces. The ewes must have thick fleeces and long staple, must be well covered on head and legs, and evenly fleeced on all parts of the body.

Know that they are well bred ; scrub stock is no good to breed. Do not think because it seems cheap that you can breed from it and improve it. It is far more profitable to start right with good stock and improve that ; improvement of good stock is a distinct profit.

SHEEPLETS.

Blood will tell.

Get good mothers and good milkers.

Uniformity is of prime importance in breeding.

Good blood is the foundation of a good animal.

Get as good quality as you can, but have the whole flock the same quality.

Never buy a poor sheep ; let the next man fool with it, you have not time nor feed to throw away, you do not need to make an experiment.

The profitable sheep are those that keep their heads close to their business of eating ; they lose no time, for they always put in good work at the trough and the rack.

Do you *know* if there is any kind of farm stock so profitable as a little flock of good sheep well kept ? I don't. Try a few sheep and see if they don't come out " on top."

CHAPTER VII

BREEDING SEASON.

Fat ewes do not raise lambs.

 The breeding season should begin about November 1st for the average sheepman, which brings the lambs about April 1st. In the Northern States experience has shown that April is the best month for lambs to come. Those who want lambs for the Christmas market, or earlier, should have their lambs come in February or March, which will necessitate breeding in September or October. But for this, a warm barn and a plenty of roots and clover hay are absolutely necessary, while early lambs require much extra care. The extra expense of raising early lambs is seldom repaid by the increased growth ; there is nothing like green grass to make milk and grow lambs.

Ewes should be bred in the fall after they are one year old, making them have lambs when two years old ; some slow-maturing ewes should not be bred until after two years old. The breeding period of ewes is every seventeen to twenty-one days. The period of gestation is calculated at one hundred and forty-five days, but five months is the average time.

Have your ewes in good thriving condition, but not fat ; that does not mean to keep them poor, for that would bring small, weak lambs. They should be grazed on an old pasture, timothy or blue-grass, for at least two weeks before breeding begins. Be careful not to use fresh clover pasture ; there is much danger of the ewes not getting with lamb. If thin in condition, a grain feed of two parts oats and one part corn, one-half pound per head, fed at night, will strengthen them.

The ram should be kept in a small place apart from the ewes during the night, and should be given a feed of grain, both night and morning, when away from the ewes. His feed should be four parts oats, one part corn, and to this add a small handful of wheat ; give about a quart to a feed.

The ram should run with the ewes during the day, and be taken away from them when they come to the barn at night. He eats and rests at night, and in the daytime the ewes are all up and stirring, so that the ram can find those in heat.

To secure the most service from a ram, the ewes should be "tried out" every morning, to determine which are in heat, and those thus found should be taken from the flock, kept in the barn, and be served during the day, allowing the ram to give each ewe but one service. In this way the strength of the ram is preserved, although requiring more time and work on the part of the shepherd. Ewes bred thus can be so marked and recorded as to enable the shepherd to tell nearly what time they will lamb. When this method is used, the ram should be kept in the barn all the time during breeding.

One ram will serve seventy-five to one hundred ewes, if well fed and properly handled. This is for mature rams in good condition. Yearlings or older should be used, as they breed stronger lambs, and can breed more ewes. Breed a large, strong, matured ram.

It is not advisable to use a ram lamb, but if not more than twenty or twenty-five ewes are to be bred, a large early lamb may be used, if the flock consists of old ewes, but not with young ewes. Such a lamb must be given special care and extra feed during service, to insure a successful lamb crop.

A good ration for breeding ewes consists of the following : ten pounds wheat bran, ten pounds corn, one pound linseed meal, mixed. This should be sufficient for twenty ordinary ewes, fed on two pounds of cut turnips, rutabagas or mangolds, each daily, divided into two feeds, morning and night. But much depends upon the roughage.

Chapter VIII

EWES BEFORE LAMBING.

After the breeding season, which should close by December 1st, the care of the ewes depends on the weather. If the winter holds open and there is no snow on the ground, let them

HORNED DORSET EWES. run out on an old blue-grass pasture as long as there is any grass. There is considerable nutriment to this grass; in fact, it is nothing but hay cured on the ground. Give them a feed of oats, mixed with a little corn and bran, once a day, and they should come to the barn at night for a mess of hay. Let them have this care so long as there is no snow. When the latter comes and covers the grass, then they must be put onto dry feed.

During the early winter plenty of fresh air and exercise, together with good rough feed, are the main essentials. The method which gives us the best result is as follows: For morning feed, a mess of grain, consisting of two parts oats and one part corn, one bushel to sixty head. For rough feed, straw and cornstalks, as much as they will eat up clean. Feed the stalks out in a field or in the yard, and also the straw. What is left will fill up the yard, and keep it clean when the weather is wet and sloppy. Always feed this rough fodder out of doors. A little brine, scattered on the straw every other morning, will make it more palatable to the sheep, and they will eat it up cleaner. At night, a good feed of clover hay is enough. Feed all they will eat, but no more ; make them eat it up clean. Mixed hay will do, but timothy is poor stuff ; good stalks are better.

Be sure to have clean, fresh water where they can have access to it at all times.

The lamb crop depends on the care of the ewes during the winter, remembering that to have healthy

COTSWOLD EWES.

lambs the ewes must be strong and vigorous, but not fat, and the main essential to this is plenty of exercise and fresh air.

The great danger in handling breeding ewes prior to lambing is that of keeping too close and feeding too highly. Ewes fed on fattening foods and kept under the same conditions as are sheep being fitted for market must not be expected to raise strong lambs. The pregnant ewe must have such food and care as will nourish and grow the lamb within her, and give it vigor and strength when it is born ; for this, muscle- and milk-forming foods and exercise are the main essentials.

The ewes which will lamb first, if you have a large bunch, should be separated from the rest, one to two or three weeks before the lambs are due to come ; put them in a warm place, and give them more and differ- ent feed. Some laxative food, such as bran and roots, should be fed with oats. One-third bran, one-third oats, and one-third roots, twice a day, in the propor- tion of about a bushel to sixty head, at a feed, makes a good grain ration.

After the ewes have lambed, the grain ration can be just about doubled, as soon as the ewes have well recovered from the effects of parturition. They should also have all the good clover hay, or mixed clover with a little timothy, that they will eat up clean.

BE CAREFUL.

The most dainty animal is the sheep.

Keep the sheep dry always on the *outside*. *Inside*, never.

Do not handle ewes heavy in lamb ; there is danger of abor- tion and of death to the ewes.

Cleanliness counts big in the management of a flock. Clean water, clean yards, and plenty of clean, sweet food.

Don't crowd your breeding ewes through some little narrow door ; crowding and jamming of pregnant ewes kills many lambs, and often the ewes.

PLATE V.

PRIZE HAMPSHIRE-DOWN EWE AND RAM.

Chapter IX.

EWES AT LAMBING.

When a ewe is preparing to give birth to her lamb, she will stand by herself, away from the rest of the flock, will refuse to eat, will look gaunt and sunken above the hip bones on either side of the backbone, and the udder becomes full. These are the first symptoms,

BROTHERS.

which generally appear from six to eighteen hours before lambing. Gradually she becomes uneasy, walking, lying down and getting up often, a passage of the water sacs soon occurs, and if everything is right, the pains should come and induce the ewe to give birth to her lamb; this latter should be soon after the passage of the water sacs, at the most not more than six hours.

If the ewe has her lamb within twenty-four hours after she begins to show signs of lambing, all is well, and Nature will do everything required; sometimes, however, especially with young ewes, a little help at the time when they are ready to lamb, helps the ewe and saves the lamb. If you happen to be present when the lamb is born, see that the skin over the nose is at once removed, to keep the lamb from smothering; if the lamb does not begin to breathe, hold his head up and blow gently into his open mouth, until he begins to breathe.

If there is something wrong, and it is impossible for the ewe to lamb, secure at once some experienced shepherd or a veterinary.

The ewes that have lambs should be separate from those that have not. The ewe will generally take care of herself and her lamb, without particular attention.

LINCOLN EWES AND LAMBS.

During and after lambing, the ewe flock need close and constant care. After the ewe has lambed, care should be taken that the afterbirth, or cleaning, is delivered, and if not within twelve hours after lambing, should be removed. In case there has been much effort on the part of the ewe during lambing, and there is danger of inflammation, take a syringe and cleanse the womb with a pint of warm water, into which has been poured five to ten drops of carbolic acid.

When the ewe has lambed, start the milk in the udder by pulling the teat between the moistened thumb and forefinger, and help the lamb to his first meal ; if the lamb is strong and lively, the ewe a good mother, and everything is all right, the lamb will help himself. Care should be taken to notice, twice a day, that the lamb is sucking both sides of the udder ; in case the ewe gives more milk than the lamb wants, be sure to suckle out the ewe with some twin or orphan lamb, and do this until the ewe's own lamb takes all her milk.

See that the ewe does not become costive and feverish ; to prevent this, feed laxative foods, such as roots and bran ; in case these do not prove sufficiently effective, and passage of the bowels is attended by effort, give the ewe two tablespoonfuls of castor oil or a similar dose of Epsom salts.

Care must be taken to withhold rich foods for a time before and after lambing, for if the milk flow be over-stimulated at this time the udder will become inflamed and distended, which may result in the loss of the ewe. Frequent and prolonged bathings with hot water will reduce the inflammation of the udder.

For the ewe that will not own her lamb, or the lamb that will not help himself, and for the lamb which is put on a ewe other than its mother, there is just one method of treatment, viz.: put in a small pen by themselves, and keep there until the lamb helps himself ; until then the lamb should be helped from five to eight times a day, the oftener the better. In most cases, three to ten days proves amply sufficient to bring the ewe to own the lamb, or the lamb to help himself, and oftentimes it does not take so long. We have found this the only practical way to make ewes own their lambs.

Sir Anthony Fitzherbert, whose "Four Books of Husbandry" were published about 1523, gives among many curious matters instructions how "To make an ewe love her lamb." If the ewe has milk and will not allow her lamb to suck, the herdsman was to put her in "a narrow place made of boards a yard wide and put the lamb to her," and if, when it tried to suckle, the ewe "smite the lamb with her head," she was to be tied with a hay rope to the side of the pen ; and if then she would not stand she was to get a little hay, and a dog was to be tied by her where she could see him, "and this wyll make her to love her lambe shortely."

At lambing time, small temporary pens, about three by five feet, should be made in or near the pen where are the flock of lambing ewes, and into one of these pens should go a ewe with twins when she has first lambed, ewes that do not own their lambs, and those whose lambs do not help themselves. Care should be taken in feeding single ewes with lambs, that they are not fed too much. Feed roots, oats and bran, but do not feed more than the ewe will eat. Also see that the ewe is watered two or three times a day ; sheep will generally drink after eating grain, and more at night than in the morning.

In rare cases the ewe may eat the tail or feet of the new-born lamb, due to fondness for her offspring, and sometimes to a lack of salt. It occurs only when the lamb is just born, as the ewe is licking him, and can be prevented by wrapping the lamb in a blanket until he is dry and able to walk, taking care that the ewe can not lick his feet. If the tail is eaten, cut it off at once.

PLATE VI.

OXFORD-DOWN RAM.

CHAPTER X

LAMBS.

The lamb is the ewe's certificate of character, and especially it tells what the character of the shepherd is.

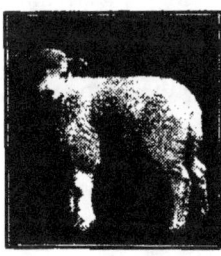 During the first year of their life, lambs should be given good feed and plenty of it, and kept growing. The quality of the flock depends on the feeding and growing of the young lambs, for a stunted lamb can never be developed into a good sheep. It is the lamb that makes the sheep, but it is the care and feed that develops the lamb into a sheep, and it is the abundance or lack of care and feed that produces a valuable and profitable animal or a scrub. Lambs are little things, and it takes only a little thing to check their growth and stunt them. Warm sheds, plenty of feed and the right kind, both for lambs and ewes, careful attention to the little details, such as seeing that each lamb finds its mother, the ewe's udder is clean, the lamb is taking all the ewe's milk, that the ewe eats well and is growing the lamb, and the many other little attentions required to keep the lamb growing, all these are necessary to convert the little lamb into a large woolly sheep that brings profit to its owner.

The ewes and lambs should be the first to go to grass in the spring and should be given the best feed. A fresh clover pasture is the best, and if this can not possibly be had, a *fresh* blue-grass pasture, but the latter should be used only a short time, in any case. Give them as newly seeded pastures as possible, and in no case put them on some old pasture, thinking any old thing good enough for sheep. Old pastures are generally productive of worms in lambs, and that means death.

For the first month after the ewes and lambs are turned to grass, have them come to the barn every

HILLSIDE PASTURE.

night ; give them a mess of grain, at first as much as they have been having on dry feed, and gradually reducing it, until by the first of June you have taken it away entirely. Start the reduction of the grain ration, first by reducing the quantity at each feed, but still giving a feed night and morning ; then take away the one in the morning, and finally the evening feed. As to hay, feed the same as when entirely on dry feed, except not so much ; feed just a very little in the morning, and turn the ewes out to pasture about nine or ten o'clock, depending on the weather. Bring them up at night, and give them a good feed of hay. In both cases, maintain the grain ration as before described.

Do not think because you have turned the sheep to grass that they can get all their living that way at once ; the change is too sudden, and your sheep will grow poor for the first month, if you give them no dry feed. Such a sudden change also often causes disease of the digestive organs, and is detrimental to the sheep in every way ; especially is this true of ewes and lambs. The grass at first is soft and flashy, and the effect is too laxative.

For early pasture, rye sown in the fall makes a very good feed, as it starts as soon as the frost is out of the ground, and grows very rapidly. For a later feed, about the last of June and July, have a field of rape to which the lambs can have access ; it is fresh, the lambs like it, and it makes them grow. Until the lambs are weaned, keep them on your best and freshest feed, and the increased growth will make itself evident in the condition of the lambs.

The lambs should run with the ewes about four to five months, which brings weaning time about August 1st to 15th. At this time put the lambs on a newly mown clover meadow from which the first hay crop has been taken. This will furnish good fresh feed ; also have the rape field handy. Then put some troughs in the lot, and once a day, either about nine o'clock in the morning or just before dark in the evening, give

WHERE ARE OUR MOTHERS?

the lambs a mess of oats and bran, two parts oats and one part bran, about a half bushel to fifty head. Always have the troughs clean, and do not feed more than they will eat up clean before leaving the troughs.

Always have clean, fresh water where the lambs can have access to it at all times. Water is half their living.

Keep salt by them at all times. Have a small box in the barn or lot which the lambs can reach but can not step into, and keep salt there all the time.

When the lambs are about six weeks old a small

ON THE ROAD TO MARKET.

pen should be fixed for them, where bran and salt should be kept all the time, and to which the lambs should have access by means of a creep—that is, a hole through which they can go and the ewes can not. This can be made by placing two boards perpendicularly, with a space between them just large enough for the lamb to go. When the lambs become used to eating bran, mix a few oats with the bran, and gradually accustom them to more oats. For this grain have a small trough at one side of the pen, where the lambs can not jump into it.

Lambs are born with an unnecessary appendage, a tail. This should be removed when the lamb is from one to three weeks old. For docking, have a man grasp the lamb firmly by the legs, holding both the right legs in his right hand and both the left legs in the left hand,

and set the lamb on his knees. The operator, with a sharp knife, disinfected with some antiseptic, removes the tail. He should find a joint in the heavy part of the tail, and aim to hit it. It will hurt the lamb less, and leave a better looking cut. If the lamb bleeds too much sear with a hot iron. The best time to do this is at night, the last thing before dark, so that the lambs can lie down, and not be disturbed or run around the pen so as to cause bleeding.

LAMB CHOPS.

In docking, do not cut off the end of the backbone.

Lambs should be weaned at four to four and one-half months old.

Sometimes the new-born lamb dies because it can not start the milk. See to this, and be sure.

Give the boy or girl a lamb and see how well they will look after your sheep while caring for their own property.

A lamb is a little thing. So is a cent. But the one goes to help make up the whole flock, and the other goes to make up the fortune of the millionaire. Care for the little things.

PLATE VII.

SHROPSHIRE YEARLING EWES.

CHAPTER XI

IN PASTURE.

Don't starve the sheep between hay and grass.

Sheep are in pasture, in the Northern States, six to eight months in the year, and it is important to have good sheep pasture. Good cattle or horse pasture is seldom good for sheep. Clover, both red and alsike, blue-grass and timothy furnish the best for sheep.

Grass should not be let to grow too high before pasturing ; sheep like short grass ; they want it sweet and fine. They need high, dry ground, and if the grass seems sparse and short it is better for the sheep; they do not like tall, coarse, rank grass.

Give them a change of pasture every three or four weeks, if it is only between two fields. Monotony kills, and the same old pasture becomes tiresome and distasteful to sheep ; it checks the growth of young stock, while the mature sheep grow thin.

Every pasture should have one or more shade trees to protect the sheep from excessive heat and storm. Another essential is cool, fresh water, which should be within easy access of the sheep at all times ; water is as important in summer as in winter.

To provide for lack of pasture due to dry weather in summer, have some rape, which can be used in connection with grass. For the lambs in the fall, rape is unequalled for green feed, but should always be used with grass. Have it adjoining the pasture, so that lambs can go to both at will.

Sow the rape in the spring, some time in May; the Dwarf Essex gives the best results; three to four pounds to the acre is seed enough. Have the ground

PURE WATER! A LITTLE SHADE! AND LOTS OF GRASS!

in fine condition, soft and mellow, sow broadcast, and cover with a weeder or light harrow. Let the rape grow to a height of twenty to twenty-four inches before pasturing; then turn in the lambs, or ewes and lambs, if the rape is ready before weaning time, let them eat it down, and then shut them off. The rape will soon grow again, and be ready for another crop of feed. Another good method is to drill the rape in rows three feet apart with a beet or garden drill and cultivate until the rape is large enough to feed off. This keeps out weeds, and cultivation makes the rape grow faster. When large enough to feed off, the sheep will walk along the rows and not trample it. For late fall feed, sow rape in the corn, just after cultivating the last time ; it will not need covering. This furnishes plenty of feed, and makes the lambs growthy and fat. It is the best supplement to grass of which we have any knowledge.

The successful fence shown in the accompanying sketch is in use on a large sheep and cattle farm; this fence was originally made of only the three barbed wires, with a ridge of earth banked up a foot high or more as a visible barrier. The sheep however crowded through between the two lower wires. The addition of the two smooth, cheap wires, one on each side of the lowest barbed wire, successfully remedied the trouble.

To make a stone wall sheep-tight use the improved method shown in the  cut. Remove a few of the top stones where the stake is to stand and incline it across the open space as shown, having the top of the stake come on the inside of the pasture wall. Replace the stones and nail boards along the upper ends of the stakes as shown. Sheep have no foothold for climbing over such a fence as this, and it is, moreover, very permanent.

Have at least one good bell for every twenty sheep. Dogs are not as likely to trouble them.

"Top-poling" a wall is not a safe fence for sheep.

They dislodge the poles, and even go over the poles, unless two in height are used. Wide boards are expensive. Foot-wide wire poultry netting can be bought for a third of a cent per running foot, with discount on large lot. Stretch it along the wall as shown in the cut. It is better than boards, and costs about one-third as much.

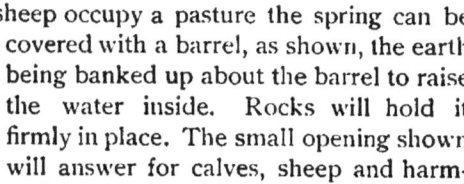

Where sheep occupy a pasture the spring can be covered with a barrel, as shown, the earth being banked up about the barrel to raise the water inside. Rocks will hold it firmly in place. The small opening shown will answer for calves, sheep and harmless animals. If used for horned animals the opening must be larger.

WEEDS.

If the weeds get the start of you on a summer fallow, or newly plowed ground, turn on the sheep, and they will eat them off, killing the weeds and feeding the sheep. For this purpose use mature sheep, either the breeding ewes after the lambs are weaned, or wethers; lambs need better feed. Oftentimes ragweeds and other noxious weeds grow up in the growing grain, especially in oats, and after harvest need trimming off, to be prevented from going to seed. Here's where your flock of sheep come in.

You want the fence corners of your grain field cleaned of grass and weeds, before sowing. That flock of ewes or wethers need just such a change for a few days. In fact, if you have short grass or weeds where you do not want them, and can let the sheep have access to them, your field is cleaned in short order, at no expense, and you are sure of a good job.

Then, perhaps you have some new ground, once chopped off, and young sprouts have grown up; if they are small enough so a sheep can reach them, turn in a flock, and see how nicely and quickly they will clean up your field for you. In fact, if you have any kind of forage on upland (sheep do not take to marsh, or wet, swampy land) that you want trimmed out, turn in the sheep and they will do it for you.

BLADES OF GRASS.

Sow turnips for the sheep.

Have good shade in hot weather.

As a manure spreader the sheep beats any modern contrivance.

Sheep keep a farm clean; they eat small brush and trim off the weeds.

Rape needs rich soil or liberal manuring, and good preparatory culture.

When sheep are on grass, feed grain at night; they won't eat it in the morning.

Sheep need all the clean, cool water they want; it is half their living in hot, dry weather.

Sheep want a variety of feed; no domestic animal relishes so much variety of feed and pasture.

Sheep running with cows seem to look upon them as

NOONTIME.

their protectors, and will run for security to the cows if dogs appear.

Do not turn to grass all at once; feed hay and grain, and let the sheep become accustomed to grass gradually.

A cold, soaking rain will take off more flesh and growth in twenty-four hours than can be replaced in twenty-four days.

Grass may be compared to milk, for it has precisely the same elements of nutrition, with about the same proportion of water.

Sheep will not graze when it is hot ; keep them in the pasture all the time during the night, in summer, and look out for dogs.

Some stockmen claim that four sheep may profitably be pastured with every cow and not materially injure the pasture for the cows.

The man who allows his sheep to get their living by pawing in the frozen snow, ought to try it himself with a pair of thin shoes on, and see how he likes it.

One of the things that count is to give the sheep a change. They soon get tired of a certain diet, quarters or pasture. Sheep will do better if changed to another pasture, though that other pasture be not quite so good.

Do not feed hay over your sheep ; have a yard to turn them in when feeding hay or straw, and shut them out of the way when feeding ; then you will keep the chaff and hayseed out of the wool, and save time in feeding.

Timothy hay is not good food for sheep when suckling lambs or before the lambs are born. Clover is best. Clover will make twice the milk, and keep the sheep from getting constipated, which is the cause of stretches and abortion. Timothy hay will dry up the milk and cause stretches.

Thousands of calves and lambs are turned out to pasture in the spring and are not housed again until fall, no protection being afforded against cold rain storms and violent showers. This results in great suffering on the part of this young stock, and a  serious check to their growth oftentimes. Trees afford a slight shelter in case of a passing shower, but for a steady rain they are worse than no shelter. Protection ought to be given, and this can be accomplished at a trifling expense in the manner shown in the accompanying cut. Four or more stakes and one length of cross pieces are all the frame required. A sloping roof of rough boards, with one side and one end boarded with the same, affords protection from cold storms.

PLATE VIII.

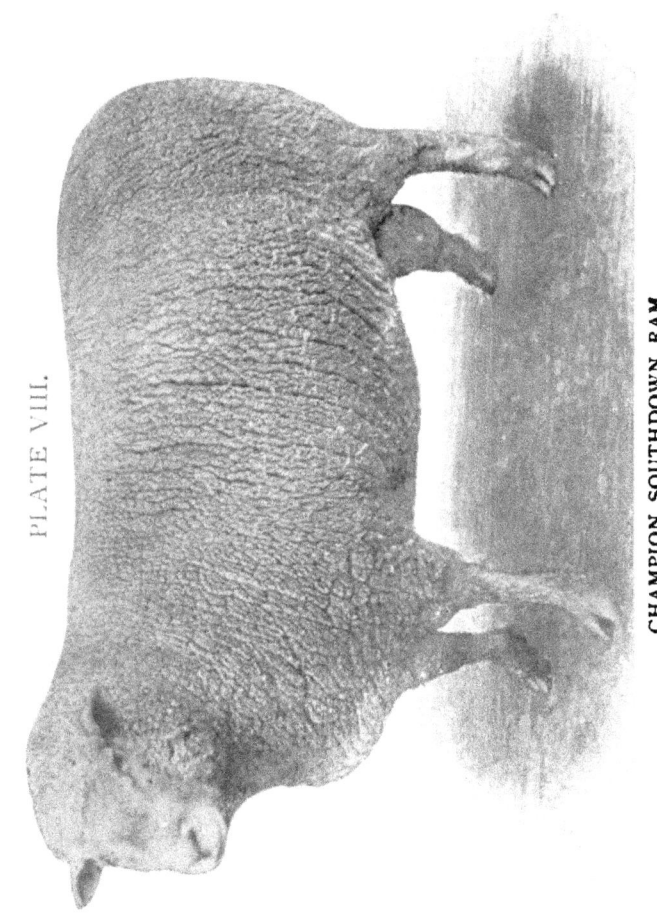

CHAMPION SOUTHDOWN RAM.

CHAPTER XII

FEEDING.

Good feed and good care are the best condition powders.

Feeding is the first essential to good stock. Cleanliness is the first essential to good feeding, and with special force does this apply to sheep feeding. The feed must be clean and fresh, and the place where it is put must be likewise. When you feed grain to sheep, see that it is clean and free from dirt, must and filth. If mud and manure finds its way into the grain room, sweep it out, and if it should accidentally get mixed with some grain, do not try to make your sheep believe that dirt is grain, or that the grain is fit for them to eat ; they will not be convinced. You will lose just that much in the condition of your sheep.

The same with your troughs, racks and water tubs. If manure and filth gets into the grain trough, see that it is not there when you put in the grain ; if you do not, you waste your grain, for the sheep will not eat grain when once tainted. Look well to the water tub ; sheep will not drink water that has been transformed into liquid manure. Clean out the water tubs often, and see that the water is always fresh and clean. Too much importance of the necessity of cleanliness can not be impressed upon the farmer who would be a careful and successful shepherd.

To be a good feeder, you must take care that you do not feed your flock more than they will eat. As many feed too much as too little. Feed your sheep only what they will eat, and have the racks and troughs empty when the sheep have finished eating, so that when the next meal time comes your sheep are hungry, and relish their feed. Overfeeding cuts both ways ; it wastes feed and damages the condition of the sheep.

For breeding ewes, straw and cornstalks once a day, in the morning, are good, and at the same time, a feed of grain, two parts oats and one part corn, bulk, a bushel to sixty head. At night, feed all the hay, clover or mixed clover and timothy, that they will eat up clean.

A sheep grower has a flock of lambs which must be handled and disposed of to the best advantage, according to the general and individual conditions. Ordinarily, the average farmer should so breed his sheep that the ewe lambs shall be of such quality as to take the place of the old ewes, and become a part of the breeding flock. For growing lambs, oats and bran, equal parts in bulk, are the best grains. They form muscle and growth, not fat. For rough feed in winter, clover hay is the best, and is about the only feed for young growing stock.

With the other half of the increase, viz., the rams or wethers (they should be the latter), there are several ways of handling. Some keep their wethers until they are two, three or four years old, as they calculate they shear enough wool to give a good profit for running, make a large increase in weight, and are more easily fattened on cheaper feeds. For this

AROUND THE FEED TROUGHS.

purpose, sheep of strong Merino blood should be used, in order to secure enough wool to make the system profitable.

The majority, however, so handle their wether lambs as to dispose of them before or during the spring they come one year old. Some sell in the fall, immediately after weaning, to feeders who make a business of buying and feeding lambs and sheep.

Others grow their lambs and fatten them themselves, and under ordinary conditions this seems to us to be the best method. Keep the wether lambs, grow them on good fresh pasture, feed them oats and bran, before and after weaning, and as soon as the pasture becomes scanty in the fall, or the weather wet or severe, put them in the barn, and begin to feed preparatory to marketing, whether it be for the Christmas market, mid-winter, or the late spring, say April or May. Which of these times is preferable depends upon the individual conditions of the grower and feeder, and upon the general conditions of the market and prospects for the season, determined by the number of sheep being fed, the price of grain, and similar conditions upon which the market is contingent, but the time for marketing should be decided upon as soon as possible, and the sheep fed accordingly; for those for a late market should not be fed so heavily nor the same kind of feed as those to be prepared for the Christmas or mid-winter market.

For fatting lambs for market, the kind and amount of feed depends upon the length of time of feeding. If for mid-winter market, *i. e.*, January or February, lambs must have been fed grain on pasture and put into the barn on dry feed not later than November 1st. While for the Christmas market, the lambs should be forced from the time they will eat grain until sold, using the following ration as near as possible for the three months next previous to marketing :

For the first month on feed, oats and corn equal parts, one-half pound per head per day ; the second month, two parts corn and one part oats, three-quarters pound per head per day ; for the third month, three parts corn and one part oats, one pound per head, and as much more as they will eat up clean. To keep the digestive organs in good shape, add to this feed five pounds of bran to a bushel of grain about once in three days. For coarse fodder, clover hay is the best. Bean pods, bright straw or good cornstalks, shredded, can be fed in the morning during the first part of the feeding period, especially if the weather is cold and the lambs' appetites are keen.

If feeding for a late market, as last of April or May, feed one-half pound per head per day, corn and oats,

A LITTLE EXTRA FEED TO HELP OUT THE PASTURE.

equal parts, until shearing or about two months before you intend to market. Shearing should be some time in February or fore part of March, depending on the weather and on the exact time intended for marketing. If you want to feed for very late market there is no need to shear until about March, as you have that much more wool. After shearing, increase the feed to the same quantity as for early-fed lambs, in proportion to the time to be fed.

Make all increases in feed gradual, and at all times see that the lambs eat up all the grain clean. During the last month on feed the lambs should have nearly all they will eat, and no coarse fodder, but good clover hay.

The average farmer must feed his sheep such feed as he raises on the farm, and he should raise such feed as will be most profitable to feed sheep. At the same time, it is often necessary to buy some feed, especially bran and oats, also sometimes various kinds of rough fodder.

For rough fodder, bean pods make good sheep feed once a day, hay the other feed. Shredded corn

fodder inside, or cornstalks outside, make good roughage for morning feed in winter. Good, bright straw makes a good alternate with the above feeds, but hay should always be fed at night. Grain should be given in the morning.

In using grain, remember corn is heating and fattening. Barley offers a partial substitute for corn; also

FEEDING CORNSTALKS.

cull beans. Oats and bran are growing, muscle-forming feeds, and operate to keep the digestive apparatus in good running order.

Beets, either mangels or stock sugar beets, and turnips furnish succulent feed, and are especially useful for ewes with lambs ; in fact, no farmer should try to raise lambs without a few roots to feed at lambing time.

To cut these roots in a shape so that sheep can eat them, have a good root cutter with knives which cut the roots into long, slim pieces, and make them just right for sheep to eat. They can be mixed with the grain, night and morning, or can be fed alone at noon.

For young sheep, *i. e.*, lambs coming a year old, clover hay at least once a day is indispensable, while for grain, oats and bran at the rate of a half pound per head per day will keep the sheep in good, growing, thriving condition. Give them plenty of exercise and fresh air; pure, fresh water is also necessary. Especially important are all three of the latter, and can not be too strongly impressed on the mind of the shepherd.

The value of salt for sheep was demonstrated by a recent feeding experiment of three months, in which sheep fed one-half ounce per head per day sheared one and three-quarters pounds more wool and weighed four and one-half pounds more than those fed no salt.

Sheep should have salt kept by them all the time. Have a small box in the pen, placed eighteen inches above the ground, and always keep salt in it. Do not leave it empty half the time and full the other half; too much salt at one time results in scouring and general loss to the sheep, and sometimes death. In summer a very good way has been to salt twice a week, dropping it upon the ground, and sheep should always have access to cool, fresh water. Salt should never be given to sheep or any other animal unless they can secure all the water they want.

KERNELS.

A NICE BUNCH.

Variety stimulates appetite.

Breeding stock should not be fattened.

Do not feed too much grain; the same holds true of all other feed.

Grow clover and oats; they are sheep feeds "par excellence."

Feed regularly; sheep know when meal time comes as well as you do.

Keep the fatting lambs quiet; noise, if it disturbs the sheep, means loss.

Do not grind feed for sheep; they have teeth, and should do their own grinding.

Watch your sheep; see that they all eat, and that they eat up what you feed them.

Sheep should have access to water at all times; like a person, they like to mix their food with water.

Make your sheep eat up the entire feed; there is no need to waste half your feed; that takes off the profits.

A little brine on the straw twice a week will make it more palatable to the sheep, and they will eat more of it.

Sheep are as neat and particular as a person, and cleanliness is as important as the kind and amount of feed.

Keep your water tubs clean; keep all manure, straw and the like out of the water; if there is manure in it, pour it out and secure fresh.

See that your grain troughs are clean before feeding grain; manure or dirt in the troughs means wasted grain, and waste means loss.

PLATE IX.

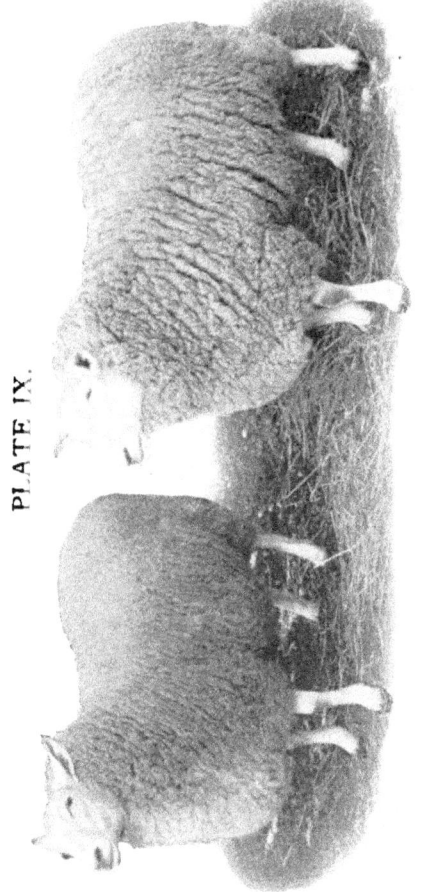

CHEVIOT RAM AND EWE.

CHAPTER XIII.

BARN QUARTERS.

However deep
The snow may sweep
The sheltered sheep
Will cosey keep.

In temperate or cold climates, barns are a necessity on the farm to preserve the feed and shelter the stock. For sheep, barn room depends on the number

A MICHIGAN SHEEP BARN.

to be cared for; if you have one hundred or more, and every farmer should have at least that many, you need a separate barn for them. However, for small farmers who can keep but few sheep, and must care for them

in a barn with other stock or in a small shed, many
of the general directions here given will also apply.
Definite and arbitrary rules for building can not be
given, because every barn must be so built as to con-
form to the particular lay of the ground and the position of the other farm buildings. An illustration of a
good sheep barn is here given.

A sheep barn should be built on a slight elevation,
with a south front, and the yard in front should slope
in the same direction ; this slope carries away all
water, and keeps both the yard and barn dry.

The barn should not be more than twenty-eight
feet wide, while the length depends on the amount of
room wanted. The barn illustrated is sixty feet long
and accommodates one hundred and twenty-five sheep,
besides having a room twelve feet wide extending
across one end, for grain, shearing, etc., while under
this room is a root cellar twelve by twenty-eight
feet, which opens into the grain room by means of a
trap door at the north end. This cellar is entirely under ground, making it warm enough so that roots will
not freeze in the coldest weather. It has stone walls
on all sides, while the floor is of the best Portland
cement. The roots are put into the cellar through a
window at the south end. The root cutter is kept at
the north end of the cellar.

Twenty-eight feet is wide enough, for the barn can
then be built with balloon frame and self-supporting
roof, which leaves the haymow free from all obstruc-
tion of any kind, and a horse fork can be run the whole
length of the mow. The track is hung in the gable, and
the hay is carried in through a door placed in one end of

the barn. Two chutes, four feet square, at opposite corners of the sheep apartment, open from the mow into a box the same size, and six feet high. These chutes make good ventilators in cold weather, keeping the air pure and fresh ; they can be closed if necessary by a trap door. Here hay is thrown down, and is ready for feeding without extra carrying or handling.

Every sheep barn should be built with eighteen to twenty feet posts, so as to have enough hay room over the sheep to supply all that can be kept in the barn through the winter. The grain room should be at one end, preferably the east, and this should open into the shed; as well as out doors. Bins  INTERIOR BARN, SHOWING ONE OF THE PENS for grain should be in one end, and the other should be used for mixing grain, for shearing, and a place in which to keep tools, drugs, wool table, and other necessary articles. It is important to have both hay and grain, and also roots, in the same barn as the sheep, as the conveniences save much time, labor and feed.

Have the barn stand the long way east and west, with the shearing room on the east end. Plenty of windows, especially on the south side, is important ; an abundance of light keeps out dampness and disease, and the air is more easily kept pure. Sheep

need light and sunshine. A basement barn does **not**
make a good sheep barn ; it is too damp, too dark
and too cold.

In the barn illustrated, water is obtained through
a hydrant connected with the tank and windmill, by
which the sheep can be furnished all the fresh water
they want, without extra work or trouble. The half
of a kerosene barrel makes a good drinking recep-
tacle. This is placed in the center of the barn, so that
all sheep in the pen can drink from the one tub. The
water is drawn from the hydrant while feeding hay.

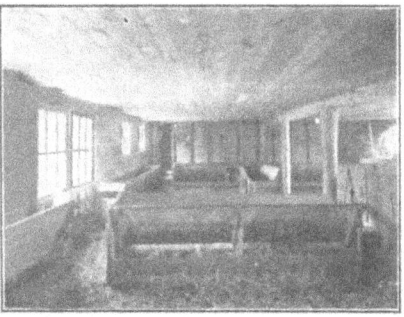

ANOTHER SIDE OF THE SAME BARN.

The floor of
a sheep barn
should be earth.
The barn should
be built on top of
the ground, and
if the soil is not
dry and porous
on top, gravel
should be drawn
in to form the
floor ; except, of
course, the grain room, which should be fitted with
matched flooring. The doors should be on rollers,
sliding back between the siding and the ceiling ; the
upper part of the doors is glass, to give light and sunshine when closed in the winter or spring. When
open, a temporary sliding gate is used in place of the
doors, to shut the sheep either way. The barn is
ceiled on the inside with matched lumber, which
makes it very warm when closed ; while the outside is

block siding. The doors should not be shut, except in very cold weather, or in case of a driving storm, or for winter lambs or shorn sheep. There are two sets of double doors on the south side, and one set on the north ; in the winter it is generally best to keep the north doors closed to prevent any draught.

SOUTHERN EXPOSURE OF SHEEP BARN, SHOWING YARD AND FEEDING TROUGHS.

In connection with the barn you must have a yard, which should be on the south side of the barn. Here you should have the grain troughs, and should always turn your sheep into this yard when feeding hay   To save time, feed the grain in the troughs before turning the sheep into the yard, and let them eat the grain while you are feeding hay. Then if you have several flocks to feed, you will not have to wait for them to eat.

Also use the yard in which to feed stalks and straw, and what the sheep do not eat will make good bedding. In any case the yard should be bedded, so as to keep it dry in the spring or in wet weather. A wet, muddy yard is a sure breeder of sore feet ; while if kept bedded, all waste straw or stalks are converted into manure, and all droppings of the sheep are saved and put back on the land.

The barn should be kept well bedded with straw, so as to preserve all the manure. This should be cleaned out of the barn once in six or eight weeks, and oftener if necessary. The barn should be kept clean, and the air pure at all times. In case it is necessary to keep closed and warm this is especially important.

The strawstack should not be where the sheep can have access to it. They pull out the straw and fill their fleeces with dirt and chaff, which damages the value of the wool at least one-third.

FRESH AIR.

Have wide doorways and gates.

Keep your feet and your sheep dry.

Give your sheep plenty of fresh air, and see that the shed is dry.

Pure, fresh air and clean, dry sheds are essential to success in the sheep business.

Sheep should not run with cattle or horses, in winter yards; there is too much danger of injury, and nothing gained by it.

All sharp corners on doors, racks, troughs, or other places should be rounded off; they waste too much wool, and injure the sheep.

Water does sheep no good, except " little and often " in the shape of drink. Outside doses make snuffles, coughs, consumption and loss.

ed, s
ather
feet
s an
,heep

:raw,
d be
-eks.
kept
nec-
ially

eep
fill
the

l is

in

ls:

es
he

ne
)n

PLATE X.

CHAMPION DORSET-HORN RAM AND EWE.

CHAPTER XIV

RACKS AND TROUGHS.

A good sheepman is always a good neighbor, or better than he would have been without sheep.

There are many different kinds of racks and troughs used for feeding sheep, while some use none at all. The latter is a most wasteful method, both as to feed and loss in the condition of stock. The racks described are those we are using, in preference to the many various patterns we have tried, and strongly recommend them for durability, practicability, ease in making, cheapness, and saving of feed.

For hay racks, we use two designs, that of the first to be preferred for such use as permits of stationary racks. Have these racks around the outside of the pen ; then, if necessary, to accommodate what sheep you have in the barn, put extra racks across the pen ; the expense of them is small.

The first pattern, as shown in the cut, can also be used for permanent partitions. The dimensions are as follows : the front of the rack is two boards, each twelve inches wide, nailed to a two-feet two by four scantling ; this two by four is bolted to the upper end of a two  by four upright, thirty-eight inches long ; to this upright, at a point eighteen inches below the bolt, nail a horizontal inch board eighteen inches long, on which to lay the bottom of the trough as shown ; this bottom must be eleven inches clear space ; the sides of this trough must be four inches clear space above the bottom; the raised portion behind the front should be six inches wide ; the front should be fastened at such an angle as to leave four to five inches space between the bottom of it and the front edge of the raised portion ; through this space the sheep pull the hay, and if any is left, it is in the trough, and not under their feet. The trough can be cleaned of all hay and used for grain if necessary.

The other pattern is thus : for uprights use two by four scantling, fifty-two inches long ; for top board, have it eight inches wide, the bottom board twelve  inches wide, while the space between the two should be eight inches ; these sides can be any length desired, but twelve or fourteen feet are the best to handle ; the ends should be twenty-four inches long ; the uprights should extend four inches below the bottom of the bottom board. This is a handy portable rack, can be placed anywhere, and is especially good to move so as to make small spaces when catching and handling sheep.

For grain troughs, we use two patterns ; the wide trough is intended for more stationary use than the V-trough and is especially good in feeding roots ; also for feeding grain to fat sheep, for they can not eat so fast, nor get dirt in the grain by jumping over the trough.

The construction of the V-trough is very simple, as follows : two boards, one inch thick, sixteen feet long, one seven inches wide, the other six inches,  the edges nailed together; for legs use two by four, twenty-four inches long, one mortised into the other at right angles, at a point seven inches from the end, leaving a place in which the trough should fit exactly; a V-shaped board nailed in each end, and the trough is complete. This is light, cheap, easily made, and handy to move from one place to another.

The wide trough is a little heavier, and costs some more to build, but for the purpose that it serves is as important as the V-trough. It is constructed thus : all lumber one inch thick, dressed ; bottom is a ten-inch board, sixteen feet long ; sides are four-inch boards, one inch longer at each end than the bottom ; the ends are thirty-two inches long, ten inches wide from the ground to the top of the side boards, from there tapering to a four-inch  width at the top ; the top should be a four-inch piece directly over the center of the trough, leaving fifteen inches between the bottom and the top, mortised into the ends ; a piece the same as the upper part of the

end boards should be put in the center of the trough
to strengthen it. To keep it from tipping over, nail a
board four inches wide edgewise to the bottom of each
end ; this board should be twenty-four inches long.

Sheep make the best use of grain when it is fed
in its original whole dry condition. Masticating their
food thoroughly, the finest weed seeds are totally
fined and destroyed. Finely ground grain forms a
sticky mass in the mouth and seems distasteful to the
sheep.

THE FIFTY-TWO POUNDS OF WOOL KIND.

PLATE XI.

PRIZE LEICESTER EWE.

Chapter XV

CARE AND HANDLING.

Sheep n e e d care nearly every day in the year ; yet many, in fact the majority of, farmers give them very little. They can live with less care and attention than any other kind of live stock, and no class of stock responds better and with greater returns than sheep when given special care, when all their humors are looked to and furnished.

The main essentials are regularity in feeding and quietness in handling. Do not scare them to death when you go among them, nor yell at them when trying to drive them or go among them.

When in a pen, keep to the outside and go slow, so as to give them time to move out of the way without tearing down the racks or hurting themselves.

When you want to handle sheep for any purpose, such as cutting feet, tagging and the like, put them in a small place where there is no chance for them to run around. Be quiet, catch them out, and put in another place when you have finished your work on each sheep. Do not chase them, but have them where they can not get away in the first place, catch them when you get ready and do not scare them. It is easy and safe if you keep quiet and learn how to do it.

To catch a sheep, grasp him by the hind leg, just above the gambrel joint, with the right hand, and pull the sheep back so that you can put the left arm under his neck. To throw down or set on its rump, let go the gambrel, grasp the flank firmly, and placing the knees against the side of the sheep, raise him off the ground and set down. In laying down pregnant ewes or very heavy sheep, instead of lifting by the flank, push the hind leg next to you off the ground and under the sheep, at the same time pushing the other side of the sheep with the other hand, the arm of which is under the sheep's neck. There is a knack in catching and throwing a sheep, which can only be acquired by practice.

In caring for sheep, it should be so arranged that the sheep have all the exercise and fresh air they want. It is a common practice to keep sheep shut in one pen through the entire winter, not even having a place out doors in which to feed grain, but it is quite wrong. Sheep are well protected with a warm coat of wool, they do not suffer from any ordinary cold weather, and they like to be out in the air. Sheep kept in a close shed on foul manure soon become dull and languid, dainty about their feed, with little appetite and no relish for what they do eat. In such a condition they will not thrive. When they exercise freely and stay in the open, fresh air, they always have a keen appetite for their feed, drink well, grow and thrive and prove profitable and easy to care for.

Every sheep barn, shed or enclosure should have an open yard in connection with it, the larger the better, and into this the sheep should be allowed to run at all times when the weather will permit. However, in extreme cold, windy weather, or in storms, especially cold rains or snows, they should have good shelter, and care should be taken that they receive it. Cold rains in spring and fall are especially to be guarded against.

See that the drinking water is always clean and fresh, that the salt box is filled, that your racks and troughs are clean, and that your sheep are all eating.

The average farmer who keeps sheep must necessarily have different ages, and those that require various kinds of care, as the breeding ewes, the growing stock, and that to be fitted for market. In handling these sheep in the various stages of growth and development, it is absolutely necessary that they be kept in separate flocks, so that each may receive the proper care, and such feed as, by the nature of the flock, is required. Young stock intended for breeders must not be fed fattening foods, nor breeding ewes given the same care as the growing lambs.

The ewe lambs should be kept by themselves, fed in such a way as to promote growth, and given all the exercise they want. Too many farmers keep the ewe lambs with the breeding ewes, which is ruinous to the lambs, if the ewes are cared for as they should be. The ewes crowd the lambs away from the feed, the feed is not such as the lambs should have, and the result is that the lambs are stunted and never make such sheep as should be put in a breeding flock.

Wethers should also be kept in a separate flock. They can be given coarser feed, and do better apart from other classes of sheep.

In sorting your flocks, preparatory to going into winter quarters, it is well, if possible, to put the older ewes, or those which may be somewhat thin in condition, into a separate flock, and give them better care and feed, and by spring they will be in good condition and ready to go with the rest of the flock.

All sorting should be done either just before going into winter quarters or in the spring before going to pasture. In winter, sheep should be kept in the same barn or shed, if possible. Changing of location and of water is always attended by a loss in the condition of the sheep, in becoming accustomed to the new location; especially is this to be guarded against in feeding sheep or lambs for market.

Trimming feet: During the winter, while sheep are kept in the barn, their hoofs grow too long, and if not trimmed the hoof turns over and lames the sheep, and in young sheep there is danger of the feet turning to one side, and producing crooked legs or ankles and impairing the value of the animal. When the feet become long, trim them with a hoof shear, like the one illustrated. Hold the sheep firmly and easily, so that it can not struggle. Cut off the hoof so that the foot will be left in its natural shape, so that the sheep can bear its

FIG. 1.

weight on the whole foot; do not cut off square, but on a slant, so that the heel and toe can both touch the ground. Notice closely how the shear is held with reference to the foot. Hoof trimming should be done in the spring before turning to grass, when the feet are the longest. The manure and lack of grit on which to wear off the shell produce long feet; wet weather and wet soil are also conducive to the growth of hoof. The proper way to hold the sheep while trimming front feet is shown in Fig. 1, and the hind feet in Fig. 2.

FIG. 2.

TAGGING : The ewes, especially the breeding flock, should be thoroughly tagged before breeding. By tagging is meant the shearing off of manure and wet, dirty wool which often adheres to the hind parts of the sheep. It is often necessary when feeding succulent foods which induce laxness of the bowels and scouring, and also when the wool on the hind parts becomes long and saturated with urine, on ewes. Besides in the fall, tagging is necessary in the summer, to prevent flies and maggots ; the sheep must be kept clean. The easiest and best way is to lay the sheep on its side, and shear off all dirty wool and taglocks, holding the sheep with its head placed over its side. Hold in an easy position, and firmly so that the animal can not struggle. Be careful not to strain ewes in lamb, by cramping, if it becomes necessary to tag them before lambing.

MARKING : Every sheepman should mark his sheep, to designate them from his neighbor's. The best way is a small metallic label in the ear, with the owner's name on one side and a series of numbers on the other. In this way the shepherd can easily keep a record of the lambs of each year, and their breeding. A common method is to use a thin mixture of Venetian red and linseed oil, dip into it a marker, wood or iron, which makes a distinctive mark, often the initials of the owner, and put on the side or back of the sheep ; the best place is on the rump, where it can easily be seen. Tar should never be used for this purpose, for it damages the wool, nor should lead paint be used ; it is as bad as tar.

THE GREATEST WEED KILLER ON EARTH.

A man starting in business requires certain articles with which to carry on his business. A farmer must have tools to till the soil ; so must he have tools and drugs to care for his sheep. Some essential articles which every farmer should always have for convenient use are the following :—a pair of sheep shears, hoof shears, a sharp knife ; a bottle of castor oil, a quantity of pulverized blue vitriol, a bottle of gasoline, in summer, a can of pine tar, a gallon of some fluid carbolic dip, a bottle of sweet spirits of niter.

To Lift a Sheep : One Person :—Catch, take the right hind leg above the gambrel in the right hand, and put the left arm around the left shoulder of the sheep, thus bringing the left hand just behind and under the fore legs ; raise and carry where you wish. Two persons, A and B :—A should put one arm under and around the neck and the other arm under the sheep just behind the fore legs ; B should put one arm under the sheep just in front of the hind legs, and the other hand on the hips, to steady the sheep and keep from kicking : or if a market sheep, B might grasp the flanks ; another simple way is for A to stand on the right side and B on the left of the sheep, A puts his right hand under the sheep just ahead of the fore legs and grasps B's left hand, then A and B lock their left and right hands respectively under the sheep just ahead of the hind legs. Both methods are simple, easy and do not hurt the sheep, not even a pregnant ewe, if done properly.

TAGS.

Keep your sheep tagged.

Wet weather induces sore feet.

Do not keep your sheep too close.

Sheep must never be allowed to get wet in winter.

Ticks on sheep make poor sheep, poor lambs and poor fleece.

In hot, wet weather do not let the maggots eat up your sheep.

Look out for lame sheep; the fouls is the forerunner of the foot-rot.

Do not compel the sheep to drink from a muddy creek or a stagnant pool.

See that your sheep have plenty of exercise; it makes them strong and healthy.

Keep your sheep's feet trimmed; in trimming be careful not to cut the toe vein.

Mutton should be more plentiful than pork because it is better and more healthful.

SCOTCH BLACK FACED RAM.

Do not catch a sheep by the wool; it is the same as grabbing a person by the hair.

Growing stock should not be kept with breeding ewes or sheep fattening for market.

Where does the poor, blue, tough mutton come from? There is a great lack of intelligence and foresight somewhere.

After weaning, the ewes should be carefully watched, and if any have a large amount of milk, they should be milked.

Warm showers will not hurt sheep, but long, cold rains do; in early spring and late fall, sheep should be sheltered from these.

Many times lambs can not be sold when they will bring the most, because the pasture is poor and they do not get fat enough.

It does not make any difference how well bred an animal you have, nor how much it cost, if you do not feed and care for it you can not expect it to thrive and be profitable.

Do not scare the life out of your sheep when you walk through the flock; go slowly and quietly, and the sheep will become so accustomed to you that they will not notice you.

A shepherd's crook is a stick with a crook at the end by which a man in charge of sheep can reach out and catch a sheep by the neck. The stick may be six or eight feet long.

PLATE XII.

TWO-YEAR-OLD LINCOLN RAM.

Chapter XVI.

DISEASES.

"An ounce of prevention is worth a pound of cure."

If sheep are given proper care and feed, and are not exposed to sudden changes, the liability of disease is materially reduced. Keep your sheep in good average condition ; do not let them get poor, nor yet keep them fat. For the average sheep that becomes sick, and you do not know how to doctor, the best way is to let nature take its course. Unless the symptoms are very evident and the remedies well known, doctoring sheep is expensive and often unsatisfactory.

In handling and treating sick animals, use common sense. Do not try to make them eat, but let them be quiet. Do not begin to pour medicine down them the first time you see there is something wrong, but look to the cause and remove it, if it is in the feed or care. If the animal does not then return to feed, study closely the symptoms, and give such treatment as the latter seem to warrant. The common ailments of sheep are comparatively few, but severe cases of many of them are very fatal.

In giving medicine to sheep, the easiest way to hold the sheep is to set it on its rump, placing the sheep between your legs and holding the head by placing the first two fingers of the left hand in the roof of the animal's mouth, thus leaving the right hand to hold

the spoon or bottle. Except where the medicine is
given clear, in one or two tablespoonfuls, the best
method is to have a long, small-necked bottle in which
to put the medicine, and put in the mouth, taking care
to put the opening well to the back of the mouth so
that the sheep can not hold the tongue over the opening. Give large doses with great care, pouring slowly
to avoid choking. Be careful not to choke by pouring
into the windpipe. In giving castor oil with a spoon,
dip the spoon in water just before using.

Keep sick animals by themselves, and do not disturb them.

A sheep is not very sick that can chew its cud.

For sheep which may be sick for several days, or
recovering from sickness and do not yet eat hay or
grain, some food is necessary, and for this nothing is
superior to a gruel, made either of flaxseed meal or
oatmeal. To make a flaxseed meal gruel, have the
water boiling, pour the meal in gradually, stirring all
the time, using enough meal to make it the consistency of a thin porridge ; let boil one minute, being
careful not to let it burn. Give one-half to one pint
at a time, by means of a long-necked bottle. *Pour
carefully, slowly.*

INTERNAL DISEASES.

CHOKING : Generally caused by too fast eating
of oats or roots, which lodge in the gullet. Set the
animal on its rump, stretch the neck and throw the
head back, and pour a cupful of water down the
throat. In more severe cases, use three or four tablespoonfuls of melted lard. If neither of these furnish

relief, take a piece of small rubber hose, or a very small, *pliable* and smooth stick, push it carefully down the gullet, and dislodge the obstacle. Keep close to the lower side of the neck, so as not to disturb the windpipe.

In passing hose to relieve choke, keep neck *perfectly* straight. Have animal held *firmly* by good assistants. Use great care to avoid wounding throat.

BLOATING : Caused by overeating of soft, green feed, such as young clover, alfalfa, rape, and the like. For slight cases, put all the pine tar possible on the nose and mouth ; also fasten a small stick in the mouth, like a bridle bit, to keep it open to allow the gas to escape. In more severe cases, give two tea-spoonfuls of bicarbonate of soda, dissolved in warm water. If relief does not follow, repeat in about ten minutes. Holding salt pork in the mouth will often relieve. In all of these cases, keep the animal in motion, so as to facilitate the escape of gas. If none of these remedies act and the animal becomes worse, tapping must be resorted to. This is done by making a small insertion with a sharp knife, at a point on the left side equidistant from the end of the last short rib and the backbone, on the paunch. Better than a knife is a trocar with shield. This is a sharp blade in a tube, and when the puncture is made the shield is left in the opening, allowing the gas to escape. This shield should be removed as soon as the animal is out of danger. Sheep trocar and canula can be secured from Jacob J. Teufel & Bro., 114 South Tenth street, Philadelphia, or other veterinary instrument makers.

FOUNDERING :   Generally caused by overeating ; for instance, securing access to grain bin accidentally, or being kept from feed twenty-four hours or longer, and then allowed to eat as much as they please.   As soon as found, give one-half teacupful of castor oil and keep well exercised.   If bloating sets in, relieve by ordinary methods.   Foundering is very dangerous, and death often results, in spite of any remedy.

CONSTIPATION :   In lambs, often occurs when one to seven days old.   Relieve by an injection, with a small syringe, of lukewarm soapsuds into the rectum.   Another good injection is glycerine, one ounce to warm water one pint.   In older sheep, sometimes due to heavy feeding, especially of corn and dry feed without any laxative foods ; also due to lack of exercise.   Two to four tablespoonfuls of castor oil will relieve ; if no passage of bowels in twenty-four hours, repeat and increase the dose by one-half.

SCOURING :   Induced by a sudden change from dry to green feed ; by overeating of green feed, such as rape, clover, alfalfa, and the like ; also of grain.   In mild cases, a change to dry feed will cause scouring to stop in a day or so, without the use of any drug. In very severe cases, where the sheep refuses to eat, and passage of dung is slimy and attended with straining, give two tablespoonfuls of castor oil to carry off the cause of the irritation ; if this does not check the passage give a tablespoonful of castor oil with thirty drops of laudanum, twice daily, in a little gruel. When checked, continue to give flaxseed gruel, until the sheep returns to its regular ration.

SNUFFLES: Similar to a cold in persons; catarrh; discharge at the nose. Put fresh pine tar in the mouth and on the nose. In severe cases steam the sheep with tar, by putting some live coals in a pan, pouring tar on them, and holding his head over the pan, placing a blanket over his head to keep the fumes from escaping, and forcing the sheep to inhale them.

URINARY TROUBLES: Rams are sometimes troubled to make water ; generally due to heavy feeding and close confinement ; it is also claimed that heavy feeding of roots will cause this trouble. Rams stand apart from the flock, do not eat, draw up their hind parts, and strain in an attempt to make water. To relieve, give one-half teaspoonful sweet spirits of niter, in a little water, every two hours until relieved.

WORMS : The deadly stomach worm (strongylus contortus) is the worst foe of the eastern sheep grower. It is a small worm about three-quarters of an inch long, found in the fourth stomach. They are taken in by lambs running on old pasture, especially blue-grass, and are induced by wet weather and wet soil ; are generally noticeable during July and August. Symptoms : lambs lag behind when driving the flock, look thin and poor, act weak, skin is very pale and bloodless ; eyes pale, sunken and lifeless ; sometimes scouring occurs a day or two before death ; death usually in four to ten days. Preventive : keep the lambs from old pastures ; a fresh cut or newly seeded clover meadow makes the best pasture ; rape is also good. Feed them some grain and dry feed, and keep some of the following mixture in the salt box all the time, viz.: one bushel salt, one pound gentian, one pound powdered copperas, one pint turpentine, mixed thoroughly. Some of the prepared medicated salts are just as cheap and effective as this mixture. Tobacco dust and tobacco leaves fed with the salt are also much used in some sections and prove very effective as a preventive. Cure: if not too bad when noticed, they can often be cured, but they are seldom as growthy as if not affected. Shut the lambs from all feed for twelve to eighteen hours ; catch the lamb, set him on his rump, holding so that he can not struggle, and give a drench of gasoline, one tablespoonful, in four ounces (one-third to one-half teacupful) of milk ; repeat the two succeeding mornings, and if no improvement, repeat the series in seven to ten days. Follow directions carefully.

———

DISEASES EXTERNAL.

MAGGOTS : Caused by green flies, induced by hot, damp weather, and dirty wool ; found on the hind part of sheep, and on rams around the horns, where wool is damp and dirty. Also around castration and docking wounds, which require watching for this trouble. Trim off the wool on place affected, and throw off the maggots ; put on gasoline to kill the maggots. Air-slacked lime will dry up the wet wool, and drive the maggots and flies away. Turpentine and kerosene are also used, but both take off the wool, if used in considerable amounts. Apply the above remedies for maggots with brush or small oil can.

FOULS, OR SORE FEET: Sheep are often lame, especially when the ground is wet; earth or manure lodges between the toes, continual rubbing induces soreness, the foot begins to suppurate, and your sheep is lame; the foot looks sore between the toes and is warm. Pare away all shell of hoof around the sore part, being sure to expose to the air all affected parts: after thoroughly paring, put on with a small swab a solution of blue vitriol and strong vinegar, mixed to the consistency of a thin paste. Keep sheep with fouls away from wet pastures or stagnant water, and keep feet dry and clean as possible.

If lame sheep are not doctored, the fouls soon spread to all parts of the foot, and foot-rot results. This becomes contagious, and all sheep remaining where are those with foot-rot will become lame. There is no need of foot-rot if the shepherd takes care of his sheep. Treat this the same as the fouls, being sure to pare away all shell and exposing the diseased parts. For a stronger solution than blue vitriol, use blue vitriol, butyr of antimony, and muriatic acid, equal parts by weight. Use with care. Paring is the principal thing; be careful not to cut the toe vein. Another excellent remedy for foul feet is one ounce chloride of zinc to one pint of water. Apply enough to wet foul parts once daily after cleaning foot with dry cloth.

TICKS: Ticks to sheep are as lice to hens; they take the life and blood from the sheep. To kill them, dip your sheep in some proprietary dip, carbolic preferred, being careful to follow directions. In winter, ticks often become a serious annoyance to sheep, especially lambs being fed for market, for sheep can not thrive and fatten when pestered by ticks ; at such times dipping is impossible, but we have found it a very good method to put the sheep in a warm barn and shear them.

Do not winter ticks ; inspect the flock well, and if ticks are found even in *small* numbers, begin *at once* to combat them. Cold weather and long wool makes treatment difficult and expensive.

SCAB : Is a strictly contagious disease of the skin, caused by a small mite which bites the skin, causes irritation to the sheep; biting and rubbing ensue, the wool is pulled out and sheds off, a crust is formed over the sore, the mites increase and spread on the edges of the sore into the unaffected parts of the skin. It generally appears on the back, rump or sides of the sheep, and is first indicated by rubbing and pulling of the wool. The disease is very contagious, common to large flocks and bands, especially on the western range. Cure : use some good proprietary dip, follow directions to the letter, dip your sheep thoroughly twice, the second dipping from six to ten days after the first, not sooner nor later than these limits. Disinfect all pens thoroughly and keep sheep from the old pastures at least two months. Scab is not very common to eastern sheep owners. Inspect all new animals at once for scab, as it is often introduced by purchasing stock ewes or rams.

SORE EYES : Caused by too much wool over the eyes, and the eyelid rolling into the eye ; also by getting something into the eye. Shear the wool away from the eye, and tie the cap of wool up off from the eyes, if necessary; if there is a film over the eye, better apply a few drops of a solution of ten grains of boric acid to the ounce of water, put in a pinch of powdered burnt alum.

SORE TEATS: The teats on ewes with lambs sometimes become sore and tender, so that the lamb can not suck. Rub twice a day with salted butter.

CAKED UDDER: Sometimes caused by weaning and not milking after the lamb is taken away. Generally occurs in heavy milkers; also occurs when lamb is still sucking, in one side of the bag at first. It is accompanied by stiffness in the hind quarters, the bag is hard, and in the first stages a thin, watery-like fluid can be drawn from the teat. Rub well and carefully, using camphorated sweet oil; the principal thing is the rubbing; try to soften the bag and keep the teat open. Many times the ewe will lose the use of that side of her bag entirely. If she does, send her to market. Where gait is stiff and udder caked, give the ewe one dram salicylate of soda three times daily for three or four days.

CASTING WITHERS: Thrusting out of the womb. It should be washed in a pint of warm water, in which has been dissolved a teaspoonful of powdered alum, and the womb replaced, and a stitch taken in the upper part of the opening of the vagina. The best way to cure such ewes is to market them or kill at once if they continue to give trouble in this respect. After replacing the womb, keep hind parts of animal quite high by standing in narrow stall made for the purpose, with floor made high behind.

GOITER : Lumps in the throat. Common to lambs when born ; also in young sheep during the first winter. Some think the latter is caused by high feeding. Apply tincture of iodine with a swab, rubbing on enough to color well the affected portion. Two or three applications, two to four days apart, should remove the worst case of goiter.

CASTRATING : Hold as for docking. Cut off a good sized portion of the end of the sac with a sharp knife, push back the sack from the testicles, grasp the latter singly, with right hand, and grasp narrow or upper portion of sac firmly with left hand, and draw out until the cord breaks. Do not cut the cord, but break it. When docking and castrating at the same time, castrate first, then dock, and release the lamb. The whole operation should not take over one to two minutes.

A CHAMPION LINCOLN WETHER.

PLATE XIII.

COTSWOLD RAM AND EWE.

Chapter XVII.

AGE.

The age of a sheep is determined by the size of the front teeth, or incisors, as they are technically known, until the animal is four years old. After that the age is determined by the general appearance.

Lambs have temporary incisors, ALL TEETH SMALL AND SAME SIZE IN A LAMB. which are short, small and narrow; permanent incisors come as the sheep matures. The first pair of the latter appear in the center of the set at about fifteen to eighteen months of age; the next pair, one on each side of the yearling teeth, at twenty-one to twenty-four months; the third pair at thirty to thirty-three months, and the fourth and last pair at about thirty-eight to forty-two months. Many sheep have a full mouth at three years old.

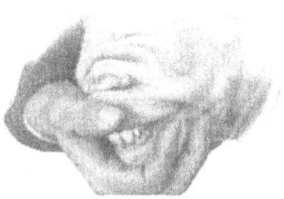

YEARLING TEETH, TWO MIDDLE LARGER THAN OTHERS.

Old sheep should not be kept. Cull your flock closely each year, marketing all ewes which are barren or that have broken mouths—that is, some or the front teeth broken. Keep the ewe lambs and sell the old ewes, always keeping young, strong ewes in the breeding flock : they raise more wool and better lambs, on less feed.

The length of a sheep's life should be limited by its usefulness. Some sheep are old and useless at five years old, and others are as profitable at eight or nine years as at three. But any sheep which has a broken mouth should go to market as soon as it can be put in salable condition. It is safe to say, however, that under ordinary conditions the ewes of the average farmer are worn out at six years old, although many are good at seven or eight, but these old ewes want too much extra feed and care to make it profitable to bother with them.

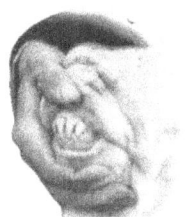

TWO YEAR OLD TEETH.

CHAPTER XVIII

PETS.

" Mary had a little lamb."

Some lamb may have lost its mother, or there may be some twin that the ewe can not raise ; let the boy, or the girl, have him to feed, and give him the lamb for his work. It gives the child an interest in his work, teaches him industry and to handle sheep.

Help him during the first two weeks. Do not overfeed the lamb. A little cow's milk, a few tea-spoonfuls at a feed, lukewarm, not cold, fed from a bottle with a rubber nipple, teaches the lamb to find his feed. Feed him often, every three hours during the day for the first week, the first thing in the morning and the last thing at night, before going to bed.

Gradually increase the amount, and after the lamb is two weeks old the child will know how to feed him. The main thing is care and regularity. Feed milk from the same cow. Feed often and regularly. Do not feed too much ; it is sure death. More pets are killed by overfeeding than any other way. Be careful that the bottle is kept clean and sweet ; rinse with cold water immediately after using, and scald with boiling water once a day. The lamb will not eat when the bottle is sour.

Have the boy raise the lamb, and when the lamb is sold let the boy have the money. He will want to raise more another year; soon he can care for the flock, and by the time he is old enough he will be a good shepherd.

The shepherd dog, or Scotch collie, is the principal herder, when sheep must be herded as on the western plains, but in the East the average farmer has little work for a dog. The sheep farmer would be better without dogs; dogs in the East are quite as much a menace to the sheep industry as are the coyotes of the West.

A FAMOUS SHEEP DOG.

Good sheep dogs need work, and that all the time, and there is not enough of it for them on a farm. There is a place for everything, but a small sheep farm has no special need of a dog.

HIS FAVORITE PETS.

CHAPTER XIX.

WOOL.

Conundrum: Why is a sheep like a government bond? Because you can take off a coupon twice a year and the bond is good as new.

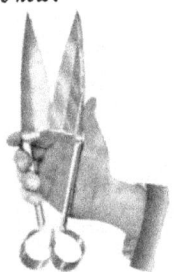

Wool is the yearly income of your sheep, which is sure, and grows with little care and expense. It is practically profit, so that the more your sheep grow and the better condition you have it in, the more money you make.

Shearing should be in April, in the northern States, generally in the last half of the month, depending on the weather. Do not shear when it is extremely cold, nor yet wait until it is so warm that the fleece

WOOL TABLE.

becomes a burden; in all cases, sheep should be shorn before turning out to grass.

The majority of sheep owners hire their shearing done by professional shearers, and to those who pursue

SHOWING HOW TOP OF TABLE RAISES

this method a few words. Have enough shearers to keep one man busy doing up the wool. Have a clean floor on which to shear, and have a pen for the sheep

opening off from it. Have one man to wait on the shearers, keep the wool out of the way and keep the floor clean. Have a table, on which to do up the wool,

FLEECE READY FOR TYING. with shorn side out. Use wool twine, but *never* sisal or manilla binding twine. Tie up in a square bundle, and put in nothing but wool. Keep out all manure, sweat-locks and dirt of every kind; also all wet tags. As to dry tags and skirts of the fleece, put them inside the fleece and tie up. Have a clean place to put your wool, some room to which rats or mice have no access, nor where dirt nor dust can accumulate.

To tie up the wool, use two strings on one side and three on the other. (See

BUNDLE READY FOR TYING.

cut of table.) After removing the fleece from the table, tie another string around the center of the fleece. For convenience in handling, leave the ends of

the outside strings of the three long enough to tie together, forming a loop, with which you can handle the fleece without tearing to pieces.

The figures 1 to 12 show

READY FOR MARKET. the consecutive positions taken in shearing a sheep. There are many variations that are permissible, according to the individual taste

of the shearer. We prefer to trim the sheep before shearing the fleece, that is, shear the legs, belly and crotch before shearing the body. This leaves the tags and skirts under the fleece, in taking up for placing on the table, the sheep is held more easily, and sits quieter than for

FIGURE 1.

trimming after the fleece is off.

After trimming, an opening is made up the under side of the neck, shearing from the left side of the sheep toward the right. Having come up the neck, shear off the cap and around the face. This done, shear around the top and about half way down the right side of the neck. When down to the left shoulder, pull the sheep over your left knee, and shear down

FIGURE 2.

the side, going to the backbone. If you want to do a smooth job, do not shear beyond the backbone, and

do not shear lengthwise of the sheep. Having shorn the left side, leaving the sheep where he sits, pull him over so you can shear the other side, assuming a position corresponding to that used in shear-

FIGURE 3.

ing the first side. In shearing, pull the hide toward you with the left hand, and do not try to pull the wool. Never mind the wool, but keep

the hide stretched smooth and taut, so that the shear can easily go into the wool.

Study carefully the illustrations, n o t i n g each position, b o t h o f t h e sheep and the shearer. The principal thing in learning to shear is keeping the sheep in position that is easy and

FIGURE 4.   comfortable for both sheep and shearer.

Do not try to take too big clips with the shears, but cut with the points, especially when first learning. There is also much in keeping the shears sharp ; dull shears will not do good work.

Shearing is something that can be learned only by practice, but beginners may be able to get some pointers from this description.

FIGURE 5.

MARKETING :   For the average sheep owner and farmer, the best method of disposing of his wool is to sell to the local dealer, of which there are generally several in every locality.   Have your wool in good, clean condition, well tied up, in a clean, light place,

FIGURE 6.   so that the buyer can see without pulling over the whole clip.

Do not put all the bad fleeces in the bottom, but

just as they come off the sheep. Do not let the buyer pull your fleeces to pieces, nor tear the whole clip so that it looks ragged.

Keep yourself posted as to the value of your wool, by reading the newspaper and reports of the mar-

FIGURE 7.

kets, so that you may be able to obtain what it is really worth. Too many farmers who raise sheep do not know what their products are worth after they raise them, because they do not read the papers and learn what others are doing and the condition of the market.

FIGURE 8.

The time to sell wool is ordinarily in the spring, soon after shearing. Wool is an article easily stored, and many fall into

the habit of holding their wool for a higher price than it is worth in the spring, thinking to secure a profit by holding, because others sell. In the long run, this is poor policy and unprofitable ; when your product is ready for market it is generally best to sell. In holding

FIGURE 9.

wool there is shrinkage, risk of holding, and other expenses which, for the average farmer, more than offset the chance of a rise in price. So that under ordinary conditions, it is well to sell when the dealer wants to buy, and be careful that you get the market price.

PELTS: If from any cause you have a sheep pelt, as soon as taken from the carcass, stretch and lay in some

FIGURE 10.

dry, clean place, flesh side up. Take wood ashes, (sifted are best), pour on hot water enough to make a good paste,

FIGURE 11.

mix thoroughly, and put on the pelt. Spread evenly all over the flesh side of the pelt, and then put one side

FIGURE 12.

of the pelt on the other, bringing the flesh sides together, with lye between. In two to six hours the wool will pull from the pelt easily; save the wool, and sell it with your next clip.

FIGURE 13.

CHAPTER XX

SLAUGHTERING.

Cleanliness is worth dollars and cents.

SHROPSHIRE WETHER.

The style in which the lamb is to be dressed depends upon the market to which it is to be sent, and even in the same market the fashion changes.

The shepherd who can dress his lambs neat and make them attractive has a great advantage over him who has to depend upon a practical butcher.

While slaughtering is an *art* and can be carried to a high state of perfection, and it is hard to learn the details from any written article, still, with a ready hand and by careful and constant study of the market, the hints herein given ought to enable the average shepherd to dress a lamb so that it will be favorably received by the most fastidious customer.

The lambs should be inspected the day before they are to be killed, weighed, and those ready for slaughter marked, so as not to disturb them unduly when they are to be caught and taken to the slaughter house. If taken out of hearing of the mothers from six to ten hours before they are to be killed, and put in a rather close, dark pen, the stomach and entrails will be more easily removed.

To kill a lamb, a trough (see Figure) should be made something like a sawbuck, with boards on

 inside of each pair of legs above where they cross, so as to hold the lamb in any position in which it is laid. It should then have its legs all tied together and be laid on left side with head extending beyond the end of trough.

By placing one hand on back of neck and other under the jaw, by a quick motion its neck can be broken. This renders the lamb unconscious and ends its sufferings.

It should then be stuck with a sharp knife just back of the ear, being sure to sever both jugular veins. Be sure to cut clean back to the vertebra, and as sure not to sever the gullet, as this would permit the contents of the stomach to escape and give a bad appearance and flavor to the meat.

" Pithing the lamb " (breaking its neck) will cause it to bleed more completely than if stuck before doing this.

Formerly New York market wanted all lambs " hog dressed," with head and feet cut off, but fashion changes. Now the feet and head are to be left on, and only a portion of the belly skinned. To do this, when the lamb is dead lay it on its back in the killing trough, rip the skin open from brisket back to a point just between the hind legs, and draw skin back a few inches on each side. Cut through the brisket, or breast bone, well down to neck. With the lamb this can be done with a heavy knife.

Now tie the hind feet together and hang upon a firm hook. Now split down the belly, being careful not to puncture the entrails. Cut the skin about the anus and carefully pull it down and remove the large intestine with the entrails. Take out with them the gullet clear down to the mouth.

The caul, the web of fat over the entrails, should be taken off and put into warm water or into a pocket made between skin and side of belly. Now insert the back-sets. These are made of any free splitting wood from twelve to fifteen inches long, one inch thick and both ends sharpened, the length depending upon size of lamb. To insert these turn the flesh of one side of belly back and stick one end of back-set through it, bring it across the back sharp, so that when both are in they shall cross on the back, and insert the other end in flesh same as the first end. Then put the other one in same way, only so as to have them cross on back and have ends three inches apart. Now take the caul while it is warm and put it on, using skewers to hold it in place until cold. Cut a small hole in the caul in proper place on each side, and having loosened up the kidneys from the back pull each one through the hole in caul, and if necessary to make it stay there skewer until cold.

All edges of the skin should be kept turned back so the wool does not come in contact with the exposed flesh, as it is liable to make it taste if it does.

For New York market merely wrap a piece of white muslin over all exposed parts, and then sew the lamb up in a piece of burlap.

The object of leaving the entire skin on the carcass is that it carries much nicer, and when the skin
is removed in the market the lamb shows very much
brighter than when sent dressed.

For other markets, Boston especially, lambs need
to be dressed very much like old sheep, or rather
undressed, for the entire skin must be taken off. To
do this neatly and expeditiously requires more skill and
practice than the New York way.

After killing the lamb or sheep, as described, lay
the carcass on its back in the killing trough and it will
be ready for "legging." "Legging" is done by
taking a leg in hand and pinching up the skin about
half way down from foot to knee or gambrel joint, and
with the knife cutting a narrow strip from the point
down to foot and cutting off the foot. Do this to each
foot, then take a fore leg between the knees and open
the skin from the point where cut down, to the center
of neck and to center of lower jaw. In opening the skin,
keep knife clean by dipping often in clean water and
wiping on clean cloth, and be careful not to cut into
meat. Let the knife go down sideways. After the
skin is opened and each side is started a little with the
knife, it can be pounded off with knife handle or piece
of clean cloth held in hand.

Treat other fore leg in same manner. Then take
hind leg and open skin down to the middle of belly
and down the belly to the point where incisions of fore
legs come together, also to the tail. Skin the hind
legs down to gambrel joints. After this the animal
must be "wizzled." "Wizzling" consists in making
incision in the neck from breast bone to jaw and tak-

ing out the esophagus or tube through which the food passes to stomach, tying it, so matter from the stomach can not escape to soil the meat. The less of the animal skinned before hanging up the better, and care should be used to prevent the wool coming in contact with the meat, as it imparts a woolly flavor that is very objectionable.

The animal is now ready to hang up. To do this, don't use the huge gambrel, but tie the hind legs with a strong cord through the gambrel tendons. The skin should now be removed from the body, and can best be done, in all parts not fastened as with a ligament, by pounding, or with a clean cloth over the hand.

A clean cloth should always be at hand and every drop of blood removed as soon as it shows so as to avoid the necessity of washing.

When the skin is removed and the head cut off, the animal will be ready for "gutting." In "gutting," first split down the belly from tail to breast bone, being careful not to cut the entrails. Now divide or split the hams apart, cut about the anus and remove the entrails entire, including the gullet or food pipe.

Now cut through the brisket or breast bone with a heavy knife, saw, axe, or cleaver.

The insides removed, the market to which it is going will determine the style in which it is to be finished. Sheep are usually plain dressed for any market, but New York calls for them with two back-sets, while Boston only takes one on sheep or lambs. Sheep usually have the "haslets" removed, while lambs have them left in until hot weather makes a liability of their spoiling. "Haslets" are the lungs, liver and heart, and when removed they are usually wrapped up and sent along with the carcass.

When full dressing sheep or lambs, the caul is removed and placed inside the skin to be kept warm, and when all is completed to this point, the kidneys are loosened from the back, the caul placed on the body with suitable holes cut in it and the kidneys pulled through. If necessary to fasten the caul upon the carcass use skewers, which should be taken out when it is cold and fixed in its place.

It is very important that all drainings of the veins be wiped off as soon as they are seen.

When shipping "full-dressed" lambs to Boston market they should first be wrapped in clean, white muslin or cheese cloth and be packed two in a crate. This is to insure their arrival in prime condition, with caul and kidneys undisturbed.

In shipping "hog-dressed" lambs to the market taking such, they should be first wrapped, all exposed parts, in white cloth and the whole enclosed in burlap as a guarantee against mutilation and to insure cleanliness in transit.

FOOT NOTES.

Have two back-sets made for each lamb to be killed before you begin.

Always have towels and bucket of water handy. Keep your hands and your knife clean.

Hay, chaff or other dirt should on no account be permitted to fall on the carcass before or after it is dressed.

Slaughter in a well-lighted building; never in the open air. In the open air the flesh becomes dark or "wind colored," as dealers term it.

A good killing knife has a blade about six inches long, one inch wide, and rather narrower at point than at hilt. It should be of good material and sharp.

Washing the meat injures its appearance and keeping quality. Wipe it with a dry cloth or with a cloth wrung as dry as possible when necessary to remove any stains.

If you have the opportunity, watch a butcher or some one who has had experience operate before trying it yourself. Then read this chapter over, and experience will do the rest for you.

Make a gallows of strong pole or four by four scantling, raised six feet from floor, with a meat hook put in every two and a half feet. Let the carcasses hang on this until cool and flesh becomes firm.

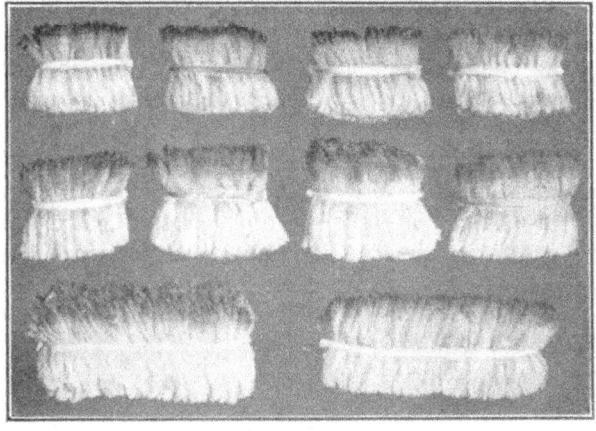

SAMPLES OF AMERICAN MERINO WOOL.

CHAPTER XXI

HOTHOUSE LAMBS.

With the right market, the right kind of man and right kind of environments, no other branch of the sheep industry will pay anything like the profit of winter lamb raising. But for any one to attempt the business who has not the proper facilities, or who is not willing to devote his most careful and constant attention to all the details, there is but one end—failure.

The folds are of great importance. For the sheep to do their best they must have light, dry, warm, well-ventilated folds and so arranged that the shepherd can give the needed care with the fewest steps.

Next in importance to the man and the folds is the selection of the ewes. Such as will produce lambs at any time and that will make them ripe and fat are not easy to find. None of the pure breeds, even the Dorset, are as good as those cross-bred. What the market demands is a medium sized but fat lamb. While the lambs of the mutton breeds will grow they will not be fat enough to bring top prices.

Undoubtedly, where land is cheap, the best ewes can be obtained by crossing the Dorset ram upon the common Merino ewe; but where a lamb, seven to ten weeks old, will sell for as many dollars, one can hardly afford to raise the ewe lambs for breeders.

Very satisfactory results may be obtained by se-

crosses, those of good bodies, low down, with good udders and from three to six years of age, and using on them Dorset rams. Lambs of this cross will be wonderfully hardy and active, and at from seven to

A CHAMPION
CROSS-BRED WETHER.

ten weeks old will be riper and with larger kidneys than any other combination.

For a few weeks previous to putting the ram with the ewes it is advisable to keep ewes on pretty poor pasture, or even to keep them in yards on dry feed. But a week before coupling change ewes to better feed and feed a little oil meal daily to cause ewes to gain. Nothing is better for this than a field of rape ; it is wonderfully stimulating. But the ewes should never be turned into a rape field when it is wet, or when their stomachs are empty, as it is liable to cause bloat and death. Better turn in an hour or two in the middle of the day, taking from another pasture and gradually accustom them to rape. Even then they should always be allowed the run of an old pasture.

The tail of a winter lamb is a sort of "trade-mark," an evidence that it is indeed a baby lamb, but it is advisable that the male be castrated. This, done when the lamb is two days old, will hardly cause any pain, and a seven weeks' old wether lamb will do better and have a larger kidney than a ram eight or nine weeks old. When the rams are left entire they are always running about teasing all the other lambs and even the ewe lambs will be so much annoyed as to be much injured.

As soon as the lambs commence to eat, which will be quite early, they should have a part of the yard fenced off into which they can go at pleasure and help themselves to all the bright, early cut clover hay they will eat. Give them fresh hay twice a day, and only ask them to eat the heads and leaves, giving the balance to other stock. Horses will eat it all the better for having been nosed over by the lambs. In addition to this, give them all the grain they will eat, cracked corn—not meal—oats, barley, wheat, peas, wheat bran, and oil cake cracked about the size of corn are all good for the lambs, and it is advisable to change often from one to another, and the troughs should be cleaned out at least twice a day, filling with fresh feed. Besides this, they should have access to all the mangolds they will eat.

There is always a good demand for a few really choice lambs for Thanksgiving, then for Christmas, and from New Year's on the market wants them in increasing numbers until the first of June, when they cease to be sold by the head and are sold by the pound, when the price often drops half in one day. The best prices are usually about the middle of February, but much depends upon the supply and demand. In good times the usual price for really fine, ripe, fat lambs runs along about ten dollars per head up to Easter, and seldom goes below seven dollars at any time for those "gilt edged."

Ewes for winter lambs should have plenty of room. To crowd them is not only bad for the ewes but for the lambs as well. A pen sixteen feet square is large enough for twenty, and to this there should be an annex six by sixteen feet for a lamb creep. The racks should be so constructed that the lambs can not get upon the feed or into the troughs in which grain is fed, and while the ewes should have all they can eat, no food should be allowed to remain in racks or troughs and get stale. It is also very necessary that they have an abundant supply of clean, fresh water always accessible. *The water dishes should be kept clean.*

A good ration for the ewes consists of clover hay, silage, corn, oats, barley, peas, wheat bran, and linseed meal. No animal loves a variety so well as sheep, and it is not advisable to confine the ewes to any steady diet.

Clover hay and silage are the standbys and should be fed every day and all they will eat of the hay and about four pounds of silage to the hundred pounds of sheep daily. It is also a good plan to feed from one to two pounds of roots—turnips to January, and mangolds after—to each, daily. Of course, this is supposing the ewes are in warm folds, for it would be foolish to expect to make greatest success with them in any other.

Of the grains, nothing is so good for the pregnant ewes as plenty of wheat bran, and this is also a fine feed for milk production. I would advise feeding it liberally at all times.

If the silage is made of corn having plenty of grain in it, the ewe will not need much corn in her ration. But it is advisable to change the grain often.

The old notion was that the ewes must not be grained much until after lambing for fear of bad results to ewe and lamb, but that is exploded. It is far better to begin graining the ewe so as to have her in good condition before the lamb drops, and if properly fed, with plenty of succulent food, no fear need be felt of bad results in lambing, or a want of milk to feed the lamb as soon as born. If the ewe be judiciously fed no trouble will be had of scouring, but if it does appear, from any error in feeding, prepare this mixture:

One fluidounce castor oil,

One dram laudanum,

One fluidounce chloric ether.

Shake well and divide into five parts, and give one to the lamb twice daily. If the lamb is very young, reduce the dose. Give in oatmeal gruel.

Chapter XXII.

THE SHEPHERD.

" The good shepherd loveth his sheep."

Success in the sheep business depends as much upon the shepherd as upon the sheep. The breed does not make so much difference, although the choice of this should be made in accordance with the location of markets, the character of the soils and the local conditions. The feeding is of importance, in that the right kind of feed, fed in proper amounts at regular times, is necessary ; but the most important of all is the man behind all this—the shepherd who does the breeding and the feeding, who looks to the every want of his sheep, who knows if a single sheep is sick or off feed, and at once seeks the cause and applies the remedy, who knows his sheep as himself, and the better the shepherd knows and does all these things, the better he likes his business and the more profit he derives from it.

The shepherd must like his sheep and care for them. Men who do not like sheep should not keep them. They can not attend to them properly and will have no interest in getting the most from their flock. Not every man is fitted to care for sheep, but there are thousands who might if they would, and knew the nature and profits of a sheep.

In the shepherd, it is the ability to notice little things, and realize their importance, that leads to suc-

cess. A sheep may not eat grain. Nine out of ten
sheep growers (they are not shepherds) pay no attention to the fact, often do not know it, the sheep
becomes poor and sickly, the grower wonders what
is the matter. Soon the sheep dies, and then the
owner becomes disgusted, and either runs down the
business or quits it. If the latter, so much the better.
But with the real shepherd it is different. He sees
that the sheep does not eat, he at once tries to find
the cause, and if it is due to overfeeding or dirty
grain, or whatever it may be, the cause is removed,
and every effort is made to bring the sheep back on
feed, and this is nearly always successful. The same
in everything connected with the care of sheep ; attention must be given to details. Sheep must eat and
drink regularly, and have clean, fresh food and water
all the time in order to thrive and be profitable. And
for all these things the shepherd is responsible.

SCRAPS.

Sheep and curs do not belong on the same farm.

Sheep dislike to have their hay racks used for hen roosts.

Do not look to the tariff for your profit so much as to the
flock.

Eighteen inches of rack room is required by each medium-
sized sheep.

Shepherds are said to be like poets, born, not made. No man
can tell, though, until he tries.

Ewes that lose their lambs should be made to raise some twin
or orphan. It is better for the ewe, and it surely helps the lamb.

You can tell a sheep farm as far as you can see by the fence
corners. There are no weeds nor briers. Everything looks clean
and thrifty.

In feeding hay or straw in racks, see that it goes in the rack, and not half of it under their feet for bedding; the latter is almost pure waste.

If our seventy-five million people eat as much mutton per head as the half as many British do, we would want fully a hundred million sheep.

Make sheep more in demand by feeding some for family use. When mutton is cured in a weak brine you can keep it and always have choice meat.

Sheep stretch and strain because they are constipated. This is called the disease of "stretches." Give bran and oil meal, and turn to grass as soon as possible.

The shed must not be close. In fact, it is better if left open on one side. Sheep need air, and it does no harm if it is cold air; but they must be protected from storms.

Sheep are the most profitable live stock kept on a farm, and there are few farms to which they are not adapted. They live on the least expensive feed, and grow two crops at the same time, wool and mutton.

THE YOUTHFUL SHEPHERD.

Chapter XXIII

THE RANGE.

About sixty per cent. of the sheep in the United States are run under range conditions, which prevail west of the Missouri River. The conditions under which sheep are handled on the range are so different from the caring of sheep on an eastern farm that it is difficult to make a farmer realize western conditions and methods.

Sheep on the range know no shelter except that of a corral, which is an uncovered yard with a high, tight pole or board fence, or of some sheltering cliff in

A WYOMING SHEEP RANGE.

time of storm. There are no barns and no sheds, but if it storms the wool is the only protection to the sheep, except as they can get into some ravine or behind some hill out of the wind.

There are no fences which turn sheep. All fencing on the range consists of three, and sometimes four, barbed wires, and are only for turning cattle or horses. Sheep are handled in flocks or bands of 2000 to 3500, generally about 2500. With each flock is a herder, who, with his dog, tends the sheep all the time, eating, sleeping and living with his sheep. He lives either in a wagon made for the purpose, containing bed, stove, provisions, and other essentials, or in a tent or cabin, depending on whether the feed for the sheep is very good, or is so scarce that it requires moving of the sheep and wagon once in two to four weeks; in the latter case a wagon is necessary.

The herder watches and protects the sheep. He directs them to where they shall feed, and if they wander too far away he turns them back with his faithful assistant, the shepherd dog. The latter is an invaluable aid in handling sheep on the range, in fact is as essential as a herder. During the day the flock spreads out and grazes, each sheep generally traveling from five to seven miles each day for his feed, picking here a bit and there a bite of the rich natural grasses that grow on the same places where once the buffalo roamed. Grass is the only feed.

At night, the sheep returning to near the wagon or tent, bunch and camp on the bedground, that is they bunch and lie down the same as a flock in a field, and require no watching by the herder through the night, unless some wild animal disturbs them or a severe storm comes up; in any case, the dog is on guard. In the morning, they are up and traveling for their feed.

Water is secured from a creek or spring, and the need of water requires the placing of the wagon near it. In winter, range which has no water in summer can be grazed, the snow taking the place of water.

Lambing is principally in April and May, and must be along some creek where there is plenty of feed and water, and if the contour of the country furnishes shelter from wind and storm, so much the better. As fast as the ewes lamb they are taken from the "drop bunch," and the ewes with lambs one to four days old are put in a flock by themselves ; as the lambs become older and stronger, the several small flocks are turned together, and when lambing is done all are in one flock again. During lambing much extra help is required, and night herding is necessary much of the time.

Shearing is done before or after lambing, according to the idea of the owner and the conditions of location, weather and feed. This is generally done at public pens, as near the railroad as possible, without necessitating too much loss of feed and flesh to the sheep ; however, large owners having 25,000 or more, often have their own shearing pens, and in connection with the latter a dipping vat. Dipping is often done soon after shearing, to kill ticks and cure scab.

The great enemies of the range sheep owner are scab and wild animals, especially coyotes and wolves. The former is cured by dipping and care in keeping from infected range ; the latter are killed in various ways, but from our knowledge, greyhounds have been quite as successful as any means tried.

Chapter XXIV.

SHEEP BREEDING ASSOCIATIONS.

The following is a list of the secretaries of the various sheep breeders' associations, which publish a flock register. There are many small and local organizations throughout the country, which can not be noticed here:

American-Delaine Merino Record, S. M. Cleaver, Delaware, Ohio.

American Cheviot Sheep Society, F. E. Dawley, Fayettesville, N. Y.

American Cotswold Association, F. W. Harding, Waukesha, Wis.

American Dorset Horn Association, M. A. Cooper, Washington, Pa.

American Hampshire Down Association, C. A. Tyler, Coldwater, Mich.

American Leicesters Breeders' Association, A. J. Temple, Cameron, Ill.

American Oxford Down Record Association, W. A. Shafor, Hamilton, Ohio.

American Rambouillet Sheep Breeders' Association, Dwight Lincoln, Milford Center, Ohio.

American Shropshire Registry Association, F. S. Springer, Springfield, Ill.

Continental Dorset Club, J. E. Wing, Mechanicsburg, Ohio.

Dickinson Delaine Record, Mrs. Beulah Miller, New Berlin, Ohio.

Improved Black Top Delaine Merino Sheep Breeders' Association, O. M. Robertson, Eaton Rapids, Mich.

Michigan Merino Sheep Breeders' Association, E. N. Ball, Hamburg, Mich.

National Lincoln Association, Bert Smith, Charlotte, Mich.

Standard Delaine Merino Record, R. M. Wood, Saline, Mich.

Vermont, New York and Ohio Sheep Breeders' Association, Wesley Bishop, Delaware, Ohio.

BIBLIOGRAPHY.

Among works to which we have had access, and which have furnished us worthy suggestions, and also which we would recommend to all farmers who, having had an interest in sheep aroused by this little work, wish to make a more extended investigation into the field of sheep literature and to learn what sheep writers of the present are saying and doing, are the following, all of which can be had of the Wilmer Atkinson Co. at the price named :

Fitting Sheep for Show Ring and Market, by W. J. Clarke, $1.50.

A Study of Breeds, by Prof. Thomas Shaw, $1.50.

Animal Breeding, by Prof. Thomas Shaw, $1.50.

Feeds and Feeding, by Prof. W. A. Henry, $2.00.

Stewart's Shepherd's Manual, $1.00.

The American Merino, by Stephen Powers, $1.50.

Domestic Sheep, by Henry Stewart, $1.50.

Modern Sheep Breeds and Management, by "Shepherd Boy," $1.50.

Sheep Farming in America, Jos. E. Wing, $1.00.

REFERENCE CHART
Showing Parts of the External Sheep.

1.	Head.	16.	Chest.
2.	Face.	17.	Shoulder.
3.	Muzzle.	18.	Elbow.
4.	Nostril.	19.	Forearm.
5.	Eye.	20.	Knee.
6.	Ear.	21.	Ankle.
7.	Cheek.	22.	Claw.
8.	Neck.	23.	Girth Measure
9.	Withers.	24.	Side or Barrel.
10.	Throat.	25.	Belly.
11.	Back.	26.	Flank.
12.	Loins.	27.	Hip Joint.
13.	Angle of Ilium.	28.	Stifle Joint.
14.	Rump.	29.	Hock Joint.
15.	Tail or Dock.		

INDEX.